VULNERABLE PEOPLE

VULNERABLE PEOPLE

A VIEW OF AMERICAN FICTION SINCE 1945

JOSEPHINE HENDIN

OXFORD UNIVERSITY PRESS
Oxford New York Toronto Melbourne

Oxford University Press
Oxford London Glasgow
New York Toronto Melbourne Wellington
Nairobi Dar es Salaam Cape Town
Kuala Lumpur Singapore Jakarta Hong Kong Tokyo
Delhi Bombay Calcutta Madras Karachi

First published by Oxford University Press, New York, 1978

First issued as an Oxford University Press paperback, 1979

Library of Congress Cataloging in Publication Data

Hendin, Josephine.
Vulnerable people.

1. American fiction—20th century—History and criticism. I. Title.
PS379.H4 813'.03 77-10103
ISBN 0-19-502319-6
ISBN 0-19-502620-9 pbk.

Printed in the United States of America

For Herbert, Erik and Neil

ACKNOWLEDGMENTS

I am grateful to the John Simon Guggenheim Memorial Foundation for their generous support of my work.

Harper's magazine originally published parts of Chapters II, III, IV, V and IX in a somewhat different form. Chapter VIII was adapted for *Ms.* and originally appeared there. *The New Republic* originally published parts of Chapter VI, and *The Nation* originally published a portion of Chapter V. I appreciate the encouragement and permission given me by these magazines.

I owe a lot to my students at the New School for Social Research for their insight and enthusiasm, and for bringing themselves so thoroughly to what they read with me.

Herbert Hendin, my husband, made this book possible through his advice, encouragement and fascination with the human situation. The themes developed in this book were the product of continual discussion between us and I am everywhere indebted to his understanding.

New York, N.Y. J.H.
December 1977

CONTENTS

VULNERABLE PEOPLE

I

IN SEARCH OF RELEASE

There is something in staying close to
men and women and looking on
them, and in the contact and odor of
them, that pleases the soul well,
All things please the soul, but these
please the soul well.

Walt Whitman
"I Sing the Body Electric,"
Leaves of Grass

Imagination and memory are but one thing, which for divers considerations hath divers names.

Thomas Hobbes, *Leviathan*

Our recent fiction has been condemned as merely a record of spiritual and emotional impoverishment. But I have found in it the richness, excitement and hope of American experience during the past thirty years. Unlike the morning newspaper, novels do not simply report our dislocations, they show how we withstand them. Even as they may expose the searing pressures in our lives, they invent their own correctives and mirror ours.

The characters in our fiction are, in a way, us. To see them clearly, to put our fiction in perspective, is to move toward putting our lives in perspective. From my student at Yale who said Vonnegut's Billy Pilgrim made him understand his passivity, to the young woman lawyer I taught at the New School who made a needlepoint sampler of Thomas Pynchon's watchword "Keep cool, but care," I have found teaching fiction has meant knowing people who are coming to novels for knowledge of themselves. The novel as case, the case as novel intrigues an audience that sees life as both problematic and exciting.

Our fiction offers the chance to see the social patterning of our personal lives. The novels discussed in the chapters that follow provide insight into what is happening to us, and invite our harshest awareness of the odds against our happiness.

It is unfashionable to think that art has positive social value, perhaps because we associate art that does with the rigid ideological stereotypes to be found in the art of totalitarian countries that repress artists. But the determined unorthodoxy of our own fiction offers a vast perspective that can be the source of a variety of correctives. Our novels reflect the many American voices and cultures ranging from the underground world of William Burroughs, to the popular culture Kurt Vonnegut seems to invent anew, to feminist protest fiction, to the highbrow classiness of John Barth and Thomas Pynchon. They represent a broad spectrum of our possible lives, enlarged and exaggerated to dramatic intensity.

Although the facts of the human drama are not new and may be unalterable, there is immense variety in how we see those facts and how we deal with them. The changes in our fiction reflect changes in our adaptation to life. The violence, coldness, sense of meaninglessness and sexual confusion that mark our fiction are widely condemned as signs of the novel's loss of power and decline as art. I believe that they serve an integrative function in the novel, or provide armor for the character, and may reflect our extreme efforts to reverse our relation to ourselves. Not all the solutions of either the novelist or the reader work; nor are all the changes that we experience good. Yet I believe there is great value in exploring them and in defining the novel's continuing ability to provide a documentary of the flexible American mind.

The novelist is the audience's audience, the observer who puts what he sees together. He may not set out to advocate a particular way of being but in writing a novel he must make one cohere; by writing to rid himself of his demons, he inevitably provides a model of how fear can be defeated. Out of the many worlds our novelists offer, I believe that one overriding vision of America emerges. Our fiction shows middle-class Americans juggling a sense of personal crisis and possibility. It mirrors many attempts to bring the eternal craving for happiness into an alignment with real and present choices. The story of the stories fiction tells is not only about, but also against vulnerability.

I believe the fiction produced over the last thirty years mirrors a search for shock-resistant lives. The search is unconscious, instinctual, and divergent. Fiction divides over the methods that will reduce emotional vulnerability and alternates between two extremes. One is holistic, stressing the virtues of management, wholeness and reason. The other is anarchic, stressing the mystical values of self-effacement and disintegration. One aims for a solution through personal durability and turmoil-free performance. The other aims at dismantling the performer. One calls for action, the other for withdrawal. They are opposite sides of the will to minimize the frustration inherent in the human condition.

Holistic and anarchic fiction each explode the traditional concerns of the novel in directions which reflect the course of American life. Political and economic reality do not have to enter the novel as explicit subjects to be there. The trend toward the consolidation of power evident in government, in the proliferation of conglomerates and in the cold war vision of the world deadlocked between the power of the Soviet Union and the power of the United States may permeate fiction as anxiety, as an acute awareness of force as the determinant of human action and of individual powerlessness. The consolidation of power and the blessings and traumas it produced may be as crucial a focus of postwar experience as fear of nuclear war.

As images in our dreams come from the world we know, so material civilization is the wellspring of fictional possibilities. The story of American industry is for the economist to tell, but its impact on ourselves is embedded in our dreams, nightmares and our novels. Dun and Bradstreet reported that more businesses failed in 1961 than in 1933. Of the 2,657,771 businesses listed in 1956 half were worth less than $10,000 and only 5 percent were worth more than $125,000. In July, 1977, Dun and Bradstreet reported that, although businesses handling volume under $200,000 predominated, accounting for some 98 percent of the proprietorships, almost 90 percent of the partnerships, and 63 percent of the corporations, their slice of total business receipts was skimpy. Accounting for 92.6 percent of the number of businesses in the United States, they earned only 11.5 percent of total business receipts—a figure that had shrunk from 14.9 percent in 1972, and from 20.5 percent in 1967. The desire for independence, for a business of one's own, may still be strong, but the trend is overwhelmingly against the individual entrepreneur. Accelerating a climb begun several years ago, the number of businesses which failed in 1975 jumped 15 percent, setting an eight-year record.

While not statistically definitive, such facts suggest some of the experiences that can change the hopes and heroes of a particular time. As the growth of industrial capitalism after the Civil War was an inspirational force for creating as our heroes the voracious financiers and entrepreneurs of Frank Norris, Theodore Dreiser and F. Scott Fitzgerald, so changes in the size and structure of American industry since World War II have had an extraordinary impact on our imagination of ourselves.

The success of American capitalism rests on methods of factory production that created an affluence based on standardized goods and services in which most people could see themselves as middle-class. The increasing importance of big business produced not only the hero as corporate man, but a change in relationships, attitudes and hopes. Erich Fromm and others have described the extent to which the state produces the kind of people who can meet its economic needs. David Riesman's *The Lonely Crowd* (1952) described people geared not for the arrow-like drive of Dreiser's Cowperwood, but for the zig-zag mobility of the executive who transfers his way up, who is willing to be a part of a team on which no one is irreplaceable, and whose ability to live by consensus values and form disposable friendships wins him a higher rung. Riesman noted that self-merchandising in this group was almost more crucial than working well. To be liked, he implies, is not only as important as being proficient. Being liked is a form of proficiency. Corporate life institutionalized attitudes and methods of marketing and manufacture that the novelist could apply metaphorically to other aspects of human behavior.

The development of fiction into a fiction of performance and one of disintegration mirrors both our economic climate and our revolt against it. The success of American capitalism has made economic conflict as a novelistic concern virtually obsolete. In fiction our heroes have changed from the ingratiating man in the gray flannel suit to, more recently, the dropout, the drifter who sees work as only a dependent relation. The fuzziness or absence of social strife, the unclear images of society in so many novels may ironically be our most precise fictional statement of how the solidification and munificence of our economic system have produced both its power and its virtual invisibility. The giant is everywhere and nowhere; it is too big to slay and too pervasive to see. Industrial technology enters novels of personal relations not as a conscious obsession but as a presence, a dream of right behavior, of durable power.

Holistic fiction may have a symbolic origin in the value placed on mechanical efficiency, in the feeling that the product is more important than how it is made. Conversely, anarchic, disintegrative fiction may have its source in the assembly line process seen symbolically as the escape from the finished product, as a process that permits one to avoid the product by concentrating on one of its parts, and that enables one to blot out the whole with a single fragment. The individual may be seen as a fabricated thing, as in John Barth's holistic novels, or as a series of disassembled parts, as in Vonnegut's anarchic stories. Holistic and anarchic fiction show different aspects of a culture preoccupied with systems of production. In this climate the individual has become the novelist's most important product. In fiction the social drama is often overpowered by and made to seem synonymous with an individual's *cri de coeur*. The writer, like many in his audience, has gone on a forced retreat to personal relations as the last arena for the play of individual power and vulnerability.

Postwar fiction has undergone a process of increasing personalization which reflects both an exploitation of technological attitudes and a desire to narrow life to manageability. The abstract concerns of the novel—"history," "art," "politics," "self"—have been personalized into tools for individual release. Our fiction has given up on the traditional literary lament for the lost greatness of prewar world culture and tradition. It has come out of mourning to stand against the literary use of history as a reproach. For example, for Eliot and the modernist writers, the personal and historical so merged that history could be a felt nightmare from which it was impossible to escape. "The past," wrote Faulkner, "is never dead. It is not even past." To step into the mind of the modernist hero was to find either a celebration of memory or a vision of memory as the open wound of the human race, a gateway to the vast history of human loss since Eden. The stream of consciousness that reproduced the rush of recalled images across the individual mind *was* the continuous river of time itself.

Current fiction blocks the flow. It celebrates the discontinuity of people from history and society, and praises the separateness of individual experience. Vonnegut's Pilgrim, for example, learns to see past, present and future as colored stripes he can cut out at will. He can clip away the disasters and thereby destroy memory as an intensifier of pain. Our fiction treats history and memory as a series of disconnected episodes. Events in time in current novels are not linked in an inexorable causal chain, nor are they part of the continuous river of life. They are the rocks that help you keep your feet dry.

We are trying to control the present and the past by making them interchangeable. We mine history for images of ourselves. Readers and writers are increasingly ignoring the differences between one event and another. The extermination of the Jews has become a free-floating metaphor for any kind of unhappiness, no matter how banal. In a recent poem it describes the boredom of waiting on a supermarket check-out line. In an Op-Ed piece in the *New York Times* it is a symbol for the treatment of homosexuals in America today. Like the now-obsolete tee-shirts that declared in orange, blue or white, "I'm Patty Hearst," we convert personal or social tragedy into standard disgruntlement. As much as this reflects a lamentable inability to tell one kind of suffering from another, it reflects the new faith of a democratic, technological society that standardization is a form of reassurance, a kind of spiritual rest.

Ideas of what is human have changed accordingly. Sartre, for example, insisted to war-wrecked France that man could create himself by choice, that he was the sum of free decisions. To proclaim the power of individual will in the face of a blitzkreig is a gesture so noble it conceals the truth. It is a form of religious faith that declares, "I believe because it's unbelievable." Sartre's romantic denial of the limits on our choices overlooks human nature. "I read Sartre," John Barth's Jake Horner remarks, "but had difficulty deciding how to apply him to specific situations (How did existentialism help one decide whether to carry one's lunch to work or buy it at the factory cafeteria?)."

The often hilarious sense of mankind's limitations that fills current fiction may express our fear that dreams of greatness can be as destructive as overt malice. It restores to a central place in our awareness the actual dimensions of our lives, the limits on our choices imposed by economic necessity, responsibility and by emotional capacity. By exaggerating the smallness of our concerns, the paucity of the will, our fiction attempts to correct the romantic tendency to base affirmations of human life on denials of weakness. It sometimes embraces its own correctives too tightly, moving to the opposite extreme in making awareness of our limitations the gateway to strength. What it prizes is the kind of durability that can be won from the clearest recognition of inadequacy.

The human ideal in current fiction may be operational man, the survivor. He translates the yearning for self-transcendence that is a persistent theme in Western art into a concern for mastering the mechanics of behavior. He wishes not to be a slave of his past, his personality, his relations with others. He may not consciously look for salvation. In fact, he may define happiness wryly, as the absence of pain. He develops defenses against the involvements he fears. He can be understood in terms of the ego psychology that emphasizes the adaptive variety and power of the man under stress. Pirandello's image for the human family was six characters in search of an author, an anchor to the past. But we seem to have become an audience of millions in search of an invulnerable character.

We are everywhere told that our faults, our selves, are returnable. From Erving Goffman's sociology of personality which claims we are role-players whose parts are dictated by self-interest or assigned by circumstance, to the charm school ad for the more beautiful you you can be, we are encouraged to see ourselves as perfectible. We support the trends in popular psychology, Behavior Modification, Transactional Analysis, which rest on the assumption that authenticity and intimacy are impossible for most people to achieve; that most of our relations are inauthentic contacts which require manipulation. Role-playing balances mechanical efficiency against sensory experience and combines our twin obsessions: change and performance. We tend to feel enlarged by choosing and exchanging parts in the relations we have; we see ourselves functionally as tools.

The books we read read us. Like our heroes and heroines we do not idealize Sturm und Drang or believe that suffering is ennobling. We look for self-transcendence in experience. We often come to see life as a management problem. We apply the methods that made Detroit famous to the production of happiness. One of the things Masters and Johnson established was that just as autistic children respond better to machines than to people, so the sexually autistic adult can come alive to vibrators, films, mechanical techniques. We may not exactly evaluate life by that standard of utility, but we have helped make books on how to live, die, marry and divorce a major industry. We try to structure our relations to reduce emotional strain. Mechanizing our lives with each other seems to allay our anxiety.

A novel that works beautifully provides us with a model of how life can cohere. The techniques that are used to pull novels together mirror the means we use to keep ourselves together and to balance our warring impulses. Irony is not only a technique of fiction, but an angle of vision from which we view ourselves. More than any other device, irony bridges the distance between our sense of vulnerability and our dreams of power. Irony both points out and yokes the endless disparities between what we say and what we do, between what we want and what we get. Since irony became our reconciler, our use of it as a comic style has changed. In *Catch-22*, for example, what was funny were the troubles of a good-hearted man caught in the comic stupidity and greed of an inhumane system. There was no possible reconciliation between the lethal institution and the good guy. Heller's black humor comes from the traditional absurdist distinction between man and the universe, man and death.

But our more recent use of irony explodes these oppositions. There is not that much difference between life and death, claims the hilarious *MASH* soundtrack: "Suicide is painless, it brings on many changes. I can take or leave it if I please." There is no difference between cops and robbers, is the comic message of *Bonnie and Clyde*. We are trying to get beyond our sense of victimization by denying there are victims. We laugh not at what sharpens the fight, clarifies our oppositions, but what shows our differences as nil. *Dog Day Afternoon*, in which the bank robber is, as in *Bonnie and Clyde*, less treacherous and less criminal than the police, goes further in bringing the audience into the film as a chorus of onlookers, militants and passersby that cheer on the violence of both cops and robbers. Bill Slocum, Heller's hero of the seventies in *Something Happened*, unlike Yossarian is not an adversary but an embodiment of the spitefulness and cruelty that mark his corporation and his country club. Social and personal themes coalesce in the insistence that the System is You.

Fiction reaches for the ironic depths where we are interchangeable with whatever once seemed better or worse than ourselves. In one of Flannery O'Connor's stories, a one-legged woman atheist believes she would enjoy seducing a Bible salesman who radiates blessedness and virginity. She discovers, as he lustfully steals her wooden leg, that he is kinkier than she is. In a John Updike novel a philandering minister writes a sermon in favor of breaking the Seventh Commandment on the grounds that adultery is a sacrament. We try to reduce the sting of conscience, of hell, by claiming the sublime and profane are alike.

Such hard ironies speak directly to the audience in search of emotional economy. We seem to want to withstand the seduction of idealism, belief, or faith that an atheist and a Bible salesman can be morally different, that anyone's motives are pure, that love is possible. We cultivate cynicism like orchids. It is easier to admit the worst for fear the best will hurt more in the end. Irony neutralizes sympathies, creates a situation where all terms of all equations are laughable, equally without emotional charge. By reducing all distinctions, all affirmations, irony multiplies us. "Commitment," as one twenty-eight-year-old graduate student acidly put it, is "to prisons and insane asylums." Such ironies enable people to use pessimism as armor, to see themselves as provisionals, possibles who are never without open doors. Irony is a national attitude, the forge on which we flatten our fears, control our emotions, and try to become iron people.

It is impossible to say for sure why irony became a tool for releasing characters from moral distinctions. The ironic sense of personal crisis distinctive to postwar American fiction may be related to the mixture of plenitude and fear that marked a period of brinkmanship politics in which the economy flourished, with rising affluence and low unemployment, but increasingly rested on military production. Of every dollar spent in 1953 by the United States government eighty-eight cents went for defense. This combination of stability and danger, of material security and anxiety produced by the possibility of nuclear war, helped build irony into the fabric of American life. It is perhaps because the consolidation of economic power in America brought prosperity and was not apparently uncontainable through existing antitrust legislation that it did not become a major source of concern. The Vietnam war made the corporate structure a more popular target, but it remained too difficult for most people to become adversaries of a system that bestowed such economic blessings.

In a sense the consolidation of power and the pomise of affluence were great and benign enough in America during this period to have shaped even our revolutionary novelists. Political movements such as the peace movement, the civil rights movement and the women's movement maintained pressure for social change or economic parity that has had varying degress of success. Our novelists have stressed the extent to which America has stamped the revolutionary in its own image. They have created renegades as acquisitive, power-hungry and violent as any entrenched political boss. They face the irony that our well-being resides in an economic structure that may be out of our control. They accept reality gone out of control and move the drama of revolution outside of politics or economics. The novelist unleashes his own revolt not in any social program but in the vision of poverty as a poverty of action and sensation. The adversary is often seen less as a repressive system than as negative or limiting emotion: inhibition, guilt or even the need for intellectual order.

Our fiction could be described as a literature of revolt, but its revolution is apolitical, existential and erotic. The socially detached renegades of middle-class fiction may have intellectual roots in the sexualization of social protest that marks the insurgency of some social historians. Herbert Marcuse's blend of Marx and Freud, *Eros and Civilization,* saw capitalism as the producer of repressive relationships which reduced people to workers judged in terms of productivity. Applying the same standards to sexual relations, he saw performance as too rigidly defined and the distinction between dominant and submissive roles as involving either exploitation by the powerful or a repression that comes to feel like death. Norman O. Brown in *Life Against Death,* a more literary, mystical book, carried the sexualization of social protest further. He suggested that the body freed from ideas of performance could overcome not merely the power structure of genital sex, but fear of death. Presumably dying would appear a form of satiation and detumescence. Inhibition for Marcuse and Brown was the mark of social oppression.

The personalization of social protest is underscored by the extent to which, of all the revolutionary movements of the sixties, the ones that have survived are those for sexual change. Sex and politics have had a long and profitable life in art as analogues for each other. What has changed in our sense of ourselves can be suggested by sketching major changes in the literary meaning of sex. The treatment of sex was once subsumed, in serious literature, in the drama of our national power and vulnerability. America, the young republic, often saw sex as part of the savage wilderness the new civilization had to subdue. James Fenimore Cooper's *The Last of the Mohicans* presents the conflict: a dashing major is drawn to an intelligent, beautiful half-breed. He has nothing to say to the pretty blond English girl who faints under the slightest stress. But Cooper, who yearned to bring English civilization to his wild country, could not condone miscegenation and the incorporation of "savagery" it symbolized. He kills off the dark beauty and marries his major to the fool. Melville's Pierre has a blond wife who is all innocent sweetness. But he is wild about Isabel, the dark, intense, imaginative woman who may be his sister and is certainly his other self as a woman. Passionate, intelligent women are such threats in these novels because they stand for social disruption, the violation of basic taboos against miscegenation and incest.

That vapid snow queens triumph and dark sirens die off in American novels of the nineteenth century reflects a need to sacrifice women as individuals to the collective good. When Edgar Allan Poe (like Cooper a political conservative) wrote that the most exciting subject for poetry was the death of a beautiful woman, he expressed the desire for women to be eternal, static images of human mortality. The sheer helplessness of figuratively dead women embodied and invited the necessity for the civilized virtues of compassion and protectiveness, the charity of the strong for the weak. Anxiety over civilizing America as well as pathetic insecurity is reflected in novels in which social and sexual stability depended on women whose repressions and fear kept anarchic impulses in check.

Sex in the early-twentieth-century novel became the affair of a state obsessed not with transplanting the forms of English society, but with the appetites and ambitions of men on the rise. The sex war and the class war constituted a familiar

American equation solved brilliantly by Fitzgerald, Dreiser and others. The poor boy lusting after the rich girl wants to possess her class. As Dreiser wrote in *The Financier*, "In the pursuit of sex and money, life reaches its highest and its most artistic phase." The American novel seemed single-mindedly to align the survival of the richest with the acquisition of women in a unique brand of sexual Darwinism.

The Protestant ethic has been transferred, in recent fiction, from work to personal relations. The class war in which sex was a weapon of the social climber has all but disappeared from fiction. It seems as unfashionable in the novel for any man to want any one woman as it is to see, as the hero of *An American Tragedy* did, that success is bigger than life. In the more recent sexual materialism of the novel the question is not "Who?" but "How many?" What the hero wants to acquire is "experience," and in the process he has become more and more of a sexual drifter.

In 1953 Saul Bellow's Augie March is the upbeat adventurer who dares life and particularly womankind to provide him with a reason for stopping. His travels mirror the trouble of finding a woman great enough to invite his commitment. In 1955, Jack Kerouac's Sal Paradise is on the road looking for the people, the experience that will change his life. He is not hoping for involvement, but for self-transcendence and transformation. Thomas Pynchon's Benny Profane (1963) is consciously not interested in stopping or being transformed by a woman. He wants to be the rolling stone that nothing and no one can touch, catch or change. He remains the same, traveling down one street after another accumulating sensations.

The rise of the work ethic of sex correlates with the development of marriage as a literary symbol for every kind of political, psychological and economic bankruptcy. In literature in the fifties, the beat hero, the working man and the poet-intellectual are often brothers in revolt against male responsibility. Robert Lowell's Miltowned husband lies in the "Dionysian bed" of his nurturing mother resentful of the needs of his equally Miltowned wife (1953). John Updike's Rabbit runs from his pregnant wife and job as a vegetable-peeler salesman

(1952). Gregory Corso's fantastic would-be bridegroom hates the in-laws' question: What do you do for a living?

> O I'd live in Niagara forever! in a dark cave beneath the Falls
> I'd sit there the Mad Honeymooner
> devising ways to break marriages, a scourge of bigamy
> a saint of divorce—

What is being attacked is a life of work and responsibility—the family syndrome and the values of "maturity."

The attack on marriage has been personalized further into a debunking of intimacy. Sexual anger is a distinctive subject blossoming in fiction where there is rage between lovers, where men cannot tolerate changes in female behavior, where the vulnerability of men to women only arouses and perpetuates hostility. As the super-civilized major and the noble savages of Cooper had to give way to the industrialists, and the financiers fell to the man in the gray flannel suit, so the corporation man is giving way to the unemployed sexual drifter who sees women as adversaries or extras.

Two memorable lovers in Robert Stone's powerful *Dog Soldiers* move an audience who seem all too painfully aware of how slim can be the line between malice and affection: when the woman-hating heroin smuggler meets the love of his life he expresses his rage at women by hooking her on heroin and his love for her by giving her all she wants, free. The ecstatic experience is not sex but the drug that blots out their anger. The love between a man and a woman that represents civilization for our earlier novelists finds in Stone's novel the bitterest possible extension. Stone offers a Vietnamed America in which aggression and self-hatred have so thoroughly triumphed that tenderness no longer exists. Civilization has given way to the rip-him-off-before-he-rips-you-off law of the junkie jungle. The strongest bond is between woman as addict and man as pusher. Sex is only his dominance and her submission.

Life has traditionally been infused into the novel by the

drama of oppression, by the energy of social groups trying to restructure their lives. Women writers are breathing life into the ancient subject of sex by providing greater insights into the female journey. From funny comic satires to a female literature of disgust, protest fiction by women writers does not merely offer a victim's-eye view of sexual war, but probes what makes women need masochism. As unsentimental about feminism as about marriage, it provides us with sexual images of our time: in the process of ridding herself of her weakness, Lois Gould's transsexual heroine harnesses her aggression and miraculously becomes a man, but a hideous one. The imperfect androgyne is the most radical symbol of a new American conflict over weakness and strength. The revolt against vulnerability through blurring sexual differences is a metaphor for a larger attempt to repair our sense of impotence by blurring political, religious and intellectual distinctions.

The youth movement of the sixties has disappeared as an entity, but many of its religious, intellectual and sexual values have become the new, middle-class ideals. The term "youth movement" I take to mean the sense of insurgency that coexisted with the civil rights, women's and peace movements, and came to encompass a general, non-specific claim against the tragic, unalterable aspects of life. Counting on a continuing affluence to support the myth of ever-expanding personal freedoms, the youth movement had the effect of idealizing experimentation over commitment, detachment over involvement, and escape from vulnerability through blurring the reality of human tragedy, through distorting the differences between the sad and the comic, the weak and the strong. Its revolutionary values filtered easily into those of the dominant culture because they did not directly attack the system so much as parody it. Consider the "new," middle-class religion, virtually tailored for the young of any age.

One of the most significant voices in American Protestantism is Harvard theologian Harvey Cox, whose books offer perfect examples of fashionable dissent. As a liberated theologian, he declares in *The Secular City,* "There is a certain validity to the Marxist assertion that existentialism is a 'symptom of bourgeois decadence,' since its categories of *Angst* and vertigo seem increasingly irrelevant to those of the new epoch. In the age of the secular city, the questions with which we concern ourselves tend to be mostly functional and operational." In *God's Revolution, Man's Responsibility* he claims, "There is a secular revolution going on. . . . God is in this revolution . . . Jesus of Nazareth was the first person to challenge the unquestionable authority of the religious world view."

Cox's Christ is an activist who is not a revolutionary on behalf of heaven but a Lord of Experience in tune with the life-styles American industry has made. "Anonymity," Cox says in *The Secular City*, "represents for many people a liberating even more than a threatening phenomenon . . . anonymity can be understood theologically as Gospel vs. Law." He likes the rootlessness of urban life ("the key characteristics of Yahweh . . . are linked to his mobility"). The new theology is so devastatingly functional and operational that the practices of the marketplace slide into the values of the pulpit. The message challenges none of the complacencies of industrial society and denies any difference between the needs of the soul and the thirst for the world. It replaces Christ as an image of universal vulnerability. It reassures people by making no distinction between spiritual happiness and the pleasure conferred by status and power. It is meant not for the social outcast but for the college-educated person who can equate personal exaltation with God and practicality with goodness. It speaks directly to people like the good Christian men of John Updike's novels who are moral revisionists, or who, like the Reverend Marshfield, can feel blessed because they sin so charmingly.

The intellectual quest of postwar insurgency is to return issues of responsibility, guilt, victimization or power to a premoral, apolitical world where the only concern is need. Any human state could then be seen as a momentary role played in an ongoing drama of appetites. Communications technology offered a model for human connections based on experience rather than judgment. Marshall McLuhan provided an intellectual framework for blurring the distinctions between the body, the mystical body and electric circuitry. "Electric circuitry is an extension of the central nervous system. Media, by altering the environment, evoke in us unique ratios of sense perceptions. The extension of any one sense alters the way we think and act—the way we perceive the world. When these ratios change, men change." The goal of this change is revolution:

> Electric circuitry has overthrown the regime of time and space and pours upon us instantly and continuously the concerns of all other men. It has reconstituted dialogue on a global scale. . . . In tribal societies we are told that it is a familiar reaction, when some hideous event occurs, for some people to say, "How horrible it must be to feel like that," instead of blaming somebody for having done something horrible. This feeling is an aspect of the new mass culture we are moving into—a world of total involvement in which everybody is so profoundly involved with everybody else and in which nobody can really imagine what private guilt can be anymore.

The media in McLuhan's view make individuality obsolete and provide experiences which will ultimately eliminate logic, introspection, privacy and the mass of large and fine distinctions that have been the stuff of intellectual life. By perceiving people as roles, or points on a circuit, McLuhan eliminates the sense of painful individual differences. Everyone is so subordinate to "the media experience" or to his role in it that the question of what the media say or who determines what they say scarcely seems important. McLuhan's escape from content into the structure of communication reflects an attempt to allay the sense of individual helplessness and to deal with the feeling of anonymity by loving it. Behind it may lie the dream of a world without clearly defined victims and oppressors, the myth of togetherness without political or moral stratification or distinction. No one may have power, but everyone feels powerful as part of an external force. This kind of escape from pain over individual helplessness is at the heart of the fiction of writers who became culture heroes like Kurt Vonnegut, R. A. Heinlein, or Richard Brautigan.

The mystical revolt against vulnerability took a similar

route, attempting to avoid the traps of power and powerlessness by not recognizing that either truly exists. Zen addressed the sense of frustration by exploiting it as a teaching device: by teasing, and puzzling, its non-messages help destroy all ideas, distinctions and differences. Zen communicates such truths as: "There is no pain, no origin of pain, no stoppage of pain and no path to the stoppage of pain. . . . There is no knowledge, no ignorance, no destruction of ignorance. . . ." Propositional truth is also exploded—"I walk on foot, yet on the back of an ox am I riding. When I pass over the bridge, lo, the water floweth not; it is the bridge doth flow." Zen Master Joshu is known for his excellent reply when asked, "When a man comes to you with nothing, what do you advise?" He instantly answered, "Throw it away!"

Entry into a world without distinctions, where life seems undifferentiated experience, was the promise of the LSD cult that imprinted on Zen austerity the mark of chemical technology. Timothy Leary, with Richard Alpert and Alan Watts, extolled LSD's ability to turn the ego into a "transparent abstraction" and return us to "precellular awareness" that "bears a striking resemblance to the unfamiliar universe that physicists and biologists are trying to describe here and now." The distinctions between science and religion, between chemically induced visions and natural insight, between power and powerlessness disappear in the dream of human life as a molecular stream. Although the dream is explicit in novels of science fiction, it is also present in the fractured forms fiction takes and the uses to which these forms are put.

In the novel of the middle-class American the political and religious unrest of the sixties has given way to the persistent personal unrest of people caught between the inability to make commitments and the inability not to. The crucial concerns of the postwar period—the consolidation of power, the awareness of personal vulnerability and the diffusion of a spirit of revolt against vulnerability—do not surface in the novel in the development of new forms so much as in the increasing hegemony and readaptation of old ones. The confessional form, for example, used by writers as different from each other as the virtuoso stylist John Barth and the feminist Alix Kates Shulman reflects the tendency to see the social drama as the individual's plight and embodies our sense of life as a personal exploration.

The proliferation of narratives that fold in on themselves, of episodes broken in time and space, reflects the need to experience life without traditional organization. Nothing is now more standard than the attack on authority of any kind. Where the narratives of Dickens or Thackeray were largely about external events—the complexities of kinship, or the marriages that defined the individual in a social scheme—our narratives are increasingly about divisive inner states and feelings. Since Virginia Woolf declared that "life is not a series of gig lamps symmetrically arranged, but a luminous halo," and E. M. Forster lamented the prison of the plot, many critics have traced what the development of the montage, the impressionist stream, the literary symbol owe to Freud's uncovering of the subconscious and the influence of film techniques. But our writers are using old modernist forms for different purposes. Where the novel of Joyce, for example, aimed to find beneath the mass of undifferentiated experience an archetypal unity, the novel today often unfolds an unending atomization of personality and an ambivalence without limit. The proliferation of narratives broken in time and space, of characters who are historyless and often futureless, does not aim at uncovering the essential coherence of life, but at revolting against the determinisms of age, death and personality.

In *Civilization and Its Discontents* Freud remarked that it was a mistake to let any one thing become the solitary source of meaning in life. Now we behave as though the way to avoid getting hurt is to multiply one's involvements, to become immune to the loss of any one role, person, experience. The broken narrative is to the recent hero what the assembly line can be to the worker: a way of avoiding the whole product, the cumulative result of his days. The aim of the narrative is not to unify feeling and act but to keep them apart. Speaking to a sense of life as a breaking story, it helps control the discomfort of confusion. It permits the character not to put together the tragic aspect of experience, to leave the puzzle of his attachments disassembled, to conceal the whole picture in its parts. This fragmentation of form operates as a pain-killer. It has, paradoxically, an integrative function, both as a means of communicating a particular world view, and as a defensive maneuver.

Whether it is dreamlike, full of people who are indistinguishable from each other (as in Burroughs), or whether it foils our boredom with imaginative leaps through space and time (as in Vonnegut), or turns a period supersaturated with argument and contradiction into a funhouse-labyrinth (as in Pynchon), the broken, involuted narrative mirrors the urgency of our need to reach each other through the language and the style of multiplicity. Novels that show the disparity between feeling and act, that emphasize our conflicting impulses, may speak of our confusion but they also show a persistent, all-surviving faith in human intelligibility.

Because the story is so immediately intelligible it has, from the parable to the moral exemplum to the animal fable and fairy tale, not only provided an index of how values have changed but has also been a vehicle for transmitting values. This is particularly so of children's stories. In 1952 David Riesman found the story of Tootle the train representative of the work ethic in an "other-directed," consumer society. Tootle is a young train who merrily wanders off the track to romp in the grass and flowers. Alarmed, his teachers frighten and lure him back on the straight and narrow path where conformity will win him success as a superchief. He is taught to fear his own impulses. Our cautionary tales today reflect the switch from money to love as a source of anxiety. They do not teach kids to fear nonconformity. They teach them to fear each other, not to idealize men or women and to avoid disillusionment at the hands of the opposite sex. They explode the age-old sexual expectations and roles in fairy tales.

In the old stories, the frog, the troll, the toad is transformed into a prince when kissed by a generous woman. In the "Sesame Street" version, when Beauty kisses the Beast, she turns into one. When Prince Charming kisses Sleeping Beauty, he falls asleep too. When a poor girl is kind to an ugly troll, and the troll turns into a prince who offers to marry her, the girl wants him to do her chores instead. When the Prince finds Cinderella's glass slipper at the ball, he wants the other slipper so he can wear these shoes himself. He has no interest in taking a perfect woman away from her mop and making her his princess. Disillusioned lovers have existed since the beginning of time; one speaks with the voice of time in *Finnegans Wake* when Anna Livia Plurabelle wistfully remarks, "I thought you were a prince, but you were only a bumpkin." But what is happening now is rather different—an attempt is being made to armor a generation against great expectations. On "Captain Kangaroo," scenes of an idyllic friendship between a little boy and a little girl who delight in each other's company flash to a song full of ominous warnings that adult life destroys love, forces people to evaluate each other by social standards of eligibility. The girl sings, "When we grow up I'll be a lady, will you be an engineer? I don't want to change at all." The warning is that adulthood, marriage and its issue may mean that painful abyss between the lady and her engineer who have nothing in common but their responsibilities. Such tales lack grace and elegance. They go too far in denying the actual power of people to civilize and complete each other. But they only begin the truths of sexual conflict our adult fiction continues. I find these stories wry recognitions that the needs of men and women are often opposed, that self-interest does override tenderness between people.

The most ambitious new stories offer antidotes to life with people who are quick-change artists. In John Gardner's children's story, "The Shape Shifters of Shorm," an emperor is disturbed by people who can suddenly "stand transformed to, for instance, an owl. It was unnatural, illogical, a violation of order—though the shape shifters, it is true, did no one any damage." Yet the emperor promises anything to the man who can kill them. The attempt costs the lives not only of the shape shifters but of the people who set out to kill them. Faced with these deaths, people protest that the emperor should have left well enough alone. The woodchopper who manages to kill the shape shifters must finally change his identity himself to survive. The message is that people can endure by self-transformation, by instability, by playing parts, but that the dream of order and stability kills. The values transmitted in such a story mirror those in much adult fiction.

Because it's true is it Art? Because it's me will it last? are questions which spring out of the intimacy between readers and writers. Students tend to judge fiction against life and not other fiction. The best novels provide more than an extension of personal life or even diagnosis of a cultural condition. Fiction offers the chance to have life all ways at once: to see one's conflicts clearly, but experience them obliquely and joyously through the prism of the novel. Stendhal defined beauty as the promise of happiness. In the continual, imperfect experiment with ways of living better, the vast cast of characters in our novels offer glimpses of beauty even in pain and insecurity and in the search for what Nietzsche hoped art would provide: imaginations with which it is possible to live.

Fiction does not reach the audience in terms of superficial advice or simplistic techniques. A reader who would sneer at a how-to book will nevertheless become engrossed in a novel for self-instruction. Fiction provides models of behavior which show how people are protecting themselves. Each of the chapters that follow develops different characters' defenses against pain. Some make life worse, some merely take the edge off bad experience, others have a better chance of reshaping life. The reader who develops the ability to identify and distinguish the adaptive and maladaptive solutions in fiction develops the power to be discerning about his own experience. The novel bestows the subtlest of gifts, perspective.

Looking at the inequities of Victorian society, Matthew Arnold put his faith in education to produce a people who could guide their lives by the principles of reason and compassion. America, which comes close to realizing Arnold's dream of universal education, has produced an educated class that is itself conflicted and a force for disruption, that seems to have lost faith in its ability to influence the socioeconomic institutions that will largely control its destiny. Yet the audience's and the novelist's emphasis on personal life may not mean simply an abandonment of positive, socially constructive impulses; it may be an extension of those impulses into the existential problems that were once the exclusive concerns of psychoanalysis and theology. Novels are imaginative projections of how we are reopening the relations between people, the play of impulse and mind, of morals and values as experiential questions.

Recent fiction has dramatized for our time the holistic and anarchic impulses persistent in the American mind. The community of spirits rigorously stratified by the Calvinist God defined the Pilgrim's idea of good. The union of states based on reason and law that grew from the Constitutional Convention, the affirmation of democratic faith embedded in New Deal legislation, marked peaks in our persistent tendency to identify happiness with social order. The individualism of Emerson and Thoreau emphasized an opposite sense of fulfillment in the identification of man with nature or God. Walt Whitman thought that the states could be pulled together by poets who could produce a universally gripping idea. What cuts across boundaries of genre, style and sensibility is a widening sense of change in our conception of ourselves. The novel, that unification of human experience, is patterning our will to order and our will to anarchy into a single search for happiness. The will to live differently is reflected in fiction not in any political idea, but in the exploration of different ways of dealing with personal experience.

I have tried to explore in depth novelists who represent different ways of perceiving and dealing with what seem to be the dominant personal concerns of the time. There are contemporary writers I admire who are not discussed because an all-inclusive survey of the period was not my intention. The chapters that follow see representative novels as art and as dramatic explorations of how we live. They are analyses of fiction as it applies to life; they are criticisms of our experience. What fiction provides is an imaginative revelation of the essential purposiveness of human character, its immense adaptive richness, its tenacious pursuit of a better or less painful life. The solutions of fiction do not always work well in life. But those that do prove one great purpose of fiction is to return us to the world, better informed.

II

THE WRITER AS CULTURE HERO, THE FATHER AS SON

> The will is one of the principal organs of belief; not because it forms belief, but because things are true or false according to the side from which they are viewed. The will which likes one side better than the other, dissuades the mind from considering the qualities of those which it does not care to see; and thus the mind, walking abreast of the will, stops to observe the aspect which pleases the will, and judges of the thing by what it sees there.
>
> Pascal, *Pensées*

Kurt Vonnegut gave the latest fatherly advice when he told a group of college graduates that Shakespeare was wrong to say "The smallest worm will turn being trodden on." "I have to tell you that a worm can be stepped on in such a way that it can't possibly turn after you remove your foot." Vonnegut's authority derives not only from his talents as a novelist, but from his vulnerability. He speaks out in his anarchic, disintegrative fiction for our incapacities and against the code of performance, achievement and endurance. Like a battered survivor he offers his defenses against feeling bad about getting crushed.

Vonnegut's art resonates with the leveling process now rampant in America, the egalitarianism that is not the result of realized opportunities but of decreasing possibilities, shared emotional and economic limitations. Hard work, intelligence, perseverance do not guarantee anything in an America oversaturated with graduates. And like the middle-aged Pontiac dealer in *Breakfast of Champions* who despairs for years over his homosexual son and comes home to find his wife has committed suicide, the audience knows that being a devoted husband and father does not guarantee satisfaction. Vonnegut's hardcover sales—which rose from only 20,000 copies of *Cat's Cradle* to over 155,000 for *Slaughterhouse-Five,* to over 500,000 for *Breakfast of Champions*—indicate that he is being bought not only by the young, whose paperback purchases of his novels also soar, but by people as old as his heroes. His insight transcends generations in its statement of the shared helplessness of fathers and sons. This powerlessness is the pot in which many male hopes for power and control melt down to the same frustration. *E pluribus unum!*

Vonnegut is the hero of the American male under siege. How else to explain the coexistence in his work of the uneventful dreariness of the average man and the wildest flights through space and time? Vonnegut is a double-threat hero who provides an accurate statement of both the trivial surface of American life and the charged innards of male fantasy. He uses science fiction as the vehicle for emotional truth, for the large frustrations contained in the small cliché, the pain canned in the aluminum-sided house.

Vonnegut is an instinctive pop artist. He makes life easy. He makes hard times fun. He packages the downs of the audience in dazzling tales. He tenderly reproduces the talk of car dealers, of optometrists whose astonishing verbal flatness is their strength. Through that simplicity he builds characters who are American Everymen, whose sheer banality is what saves them. Vonnegut makes of their trivia a necessary burden, an anchor that keeps them from flipping out from the force of their unhappiness. He is our hero because he has made a monument out of insecurity and turned the crushed, ground-down man into an unmistakable *model* of masculinity.

Vonnegut frankly offers himself as the voice of the many who are one with their limitations. In *Wampeters, Foma and Granfalloons,* a collection of interviews, anecdotal stories and personal odds and ends, he administers truth in doses that stop thought. His folksy cracks and peremptory style permit no development after the laugh, the snort of recognition, the encapsulating cliché. He is a master of the instant defense, the immediate retreat, the phrase that simultaneously expresses his point and blocks your disagreement. Vonnegut is the new man considering what he can count on.

Political change? "This is a conservative nation," Vonnegut

tells Wheaton College students. "It continues to do what it has always done, for good or evil. It will continue to treat non-white people badly. It has always done that. It's lazy about change. I'm lucky to be the color I am and to do what I do. This is the place for me."

Cleverness? "One of my favorite cartoons shows a couple of guys chained to an eighteen-foot cell wall, hung by their wrists. Their ankles are chained, too. Above them is a tiny barred window that a mouse couldn't crawl through. And one of the guys is saying to the other, 'Now here's my plan . . .' It strikes me as gruesome and comical that in our culture we have an expectation that a man can always solve his problems. This is so untrue it makes me want to laugh—or cry."

His parents? Vonnegut tells us he grew up in "a big brick dreamhouse designed by my architect father, where nobody was home for long periods of time, except for me and Ida Young," a black cook. He says his mother committed suicide, but he does not say when or why.

Love? "If somebody says, 'I love you' to me, I feel as though I had a pistol pointed at my head."

Solitary equanimity? Vonnegut falls into depressions. When he does, he says he can sleep eight hours at night and nap from one to five in the afternoon. "Until recently, I blew my cork about every twenty days. I thought for a long time that I had perfectly good reasons for these periodic blowups. But only recently have I realized that this has been happening regularly since I was six years old. There wasn't much the people around me could do about it." Ritalin helped him out of his depression. "I was so interested that my mood could be changed by a pill." Non-Freudian doctors are helping him deal with his anger, which, Vonnegut feels, has to do with his internal chemistry. "You know, we don't give a shit about the characters' childhoods or what happened to them yesterday—we just want to know the state of their bloodstreams."

The conversation of friends? *Playboy* interviewer: "You mean you want to be with people who live nearby and think exactly as you do?" Vonnegut: "No, that's not primitive enough. I want to be with people who don't think at all, so I won't have to think either."

What goes on in the mind of the man who is such a great talker that he knows how to keep quiet while he speaks? To find out, you have to get beyond his opinions, beyond his funny, impenetrable voice. *Situations* are Vonnegut's subvocal speech. As fabulist, as creator of the startling imaginative predicaments of science fiction, Vonnegut captures the emotions, the fantasies too painful to name, the problems the audience finds insoluble.

Fortitude is a screenplay parable riveted in a male conflict over submission and control. Young Dr. Frankenstein, shattered by the death of his mother from "cancer of the everything," devotes himself to medical technology. He spends years developing ways to hook people up to mechanical kidneys, pancreases, hearts. He finally meets his one and only patient: a beautiful older woman who looks exactly like his mother and is dying. He saves her until she is nothing but a beautiful head in a beautiful room hooked up to a basement full of mechanical devices through which he controls all her moods. Her arms are specially made so that she cannot point a gun at her head or bring poison to her lips. There is no way this woman can leave him. He has not only fixed it so that she can live 500 years, he has also arranged it so that he can be spliced into her mechanical organs. "Your kidney will be my kidney!" "Your heart will be my heart!" He gets his wish when she shoots him after discovering she cannot shoot herself. He wakes up smiling next to her, another head attached to machines, the man who lost his body and mind to supply a woman with life and himself with a mother who cannot die. He has found a no-risk woman.

This play is as strong as it is not because it is a funny, macabre attack on medical technology or on the madness of the scientist, but because it opens up a Pandora's box of male impulses. There is the sick underbelly of the urge to control, the fear of loss so intense that it overpowers any concern for the "beloved." There is the fear that the craving for dominance is at heart a hunger to be dominated, controlled, engulfed by a woman, to be simply, mindlessly nurtured by an umbilical flow of air and food. There is the vision of the horror marriage: the endless connection to a woman who cannot be pleased, whose wish to die ultimately kills her man. Like Winston Niles Rumfoord, the rich man in *Sirens of Titan* who goes off in his spaceship to escape his troubles only to encounter a force that reduces him to a string of molecules waving between the sun and Betelgeuse, Dr. Frankenstein discovers escape from his loss only in the fragmentation of his body. Rumfoord can materialize on Earth as a man only once in a couple of months. Frankenstein as lover is half machine. Is it that hard to be a man? Is the only enduring connection the gravitational pull of stars, the mechanical junction beyond life? Where does this kind of pessimism start?

In *Cat's Cradle* hopelessness is a parental gift. Vonnegut's vision of the human family radiates from two earthbound families, one white, one black, both bound together in coldness, in limitless emptiness, in a child's perception that his parents have nothing to give but ice. In the Hoenikker family, the mother died bearing her third child, but was so depressed during the lives of the first two that she might as well have been dead. The father, Felix, is the father of the A-bomb. He turns his children into walking bombs when he leaves them his last discovery, ice-9. Derived by restacking water molecules, an ice-9 chip dropped in water sets up a chain reaction that freezes all the water around it to an ice unmeltable below 114 degrees. Touched to the lips, it freezes the blood and kills; dropped into the sea it can end the world. Hoenikker's children split the chunk in three, put their chips in thermos jugs and literally acquire the destructive potential of chips off the old block.

Hoenikker can play with the properties of the physical world as though none of them were fixed. Vonnegut seems to believe that people, not molecules, are so fragmentable that even their own blood cells can be stacked against them. The Hoenikker children are stacked for misery. Angela, the oldest, who mothers the family, is a lonely, forlorn girl who dropped out of school when her mother died and does nothing but play the clarinet to records; Frank, a mechanical genius, is wanted by the police; Newt, the lovable youngest, is a midget. Ice-9 brings each of them the chance to trade his or her death-heritage for happiness. The Hoenikker children only want to be normal and whole. But through their craving to be like everybody else the world ends. Is death wholeness? Is normalcy death? Is there any alternative?

Worship your own fragmentation! Bokonon, Hoenikker's black counterpart, restacks people's emotions the way Hoenikker restacks molecules. What makes people one is their power to break apart. Bokonon juggles feelings—worthlessness, depression—into a system of phony moral opposites based on a muscle-building method, dynamic tension. The strung-out masses worship Bokonon and their own shared fragmentation. " 'God made mud,' " they chant, " 'And I was some of the mud that got to sit up and look around. Lucky me, lucky mud.' "

Bokonon and his adopted daughter, Mona, run the show of good and evil without believing in either, or even in the luck of walking mud. What Bokonon gives Mona is more than the emotional equivalent of ice-9, it is the capacity to feel nothing. When the world freezes over, Mona steps out of her shelter to look for other survivors and finds masses of ice-9 suicides arranged around her father's note explaining his advice to his flock: "God is trying to kill you. . . . Have the good manners to die." Mona bursts into "laughter, touched her finger to the ground, straightened up, and touched the finger to her lips and died."

Vonnegut's fictional children always inherit killing coldness. Hoenikker bequeathes it not merely in his invention, but in his detachment from everything but molecular games. Bokonon does it with his advice; his *involvement* in other people's misery leads him to preach death. By whichever route, Vonnegut's characters arrive at detachment and numbness as alternatives to basic betrayal. Betrayal twists the hearts of his many motherless children. The sea in *Cat's Cradle* loses its usual connotations of life and generation when it becomes the agent of death. But the frozen sea is Vonnegut's image for mother, for the mother that is dead, or so uninvolved she might as well be. The motif of the mother dead before a child's life begins, the love lost at the beginning of time, runs through Vonnegut's work, spilling into his many images of destruction and coldness.

Separation and war are the twin points of *Mother Night*. What fascinates Vonnegut in the German and American spirit is articulated by Goethe's Mephistopheles whom he quotes: "I am part of the part that first was all, part of the darkness that gave birth to light and now disputes with Mother Night her ancient rank and sway.". In Vonnegut everyone is a bright angel dimmed finally by his own origins, by the Mother Night that is not merely the void, but the void left by a human parent's withdrawal. Darkness is the maternal presence peculiarly sought by Vonnegut's beleaguered characters. Once the irreplaceable woman has been lost, everyone else seems interchangeable, replaceable. Once the irreplaceable wholeness is gone, fragmentation is all.

Howard Campbell, Vonnegut's hero in *Mother Night*, loved the wife who died before the novel begins. Their fusion in sex, the lost erotic Eden he called the "nation of two," is ended by war. War is the largest possible force for fragmentation. "The crazy loom of modern history" splits people into factions; the swastika, the star of David, the hammer and sickle, the Stars and Stripes, each offer, for Vonnegut, the "chance to go crazy in a way people find irresistible." What is always irresistible to Vonnegut's people is schiziness, what a CIA man calls the talent "to be many things at once—all sincerely—it's a gift."

Everyone in *Mother Night* is so fragmented he is both himself and his opposite: Howard Campbell, the prominent Nazi propagandist, is an American agent; Campbell's best friend, a painter, turns out to be a Russian agent who turns him in to the Israelis to be hung as a war criminal. The lovely woman who tuns up claiming to be his first wife turns out to be his sister-in-law who has become a communist agent and is involved in a plot to kidnap him to Moscow. Everyone in this novel irresistibly betrays himself and the people he says he loves. Separation and loss are the basic truths; betrayal the norm. Against this bitter truth is Vonnegut's yearning for wholeness, for fusion with the love before the war of life began. The will to whole simplicity runs through the novel, culminating in Campbell's decision to hang himself. Vonnegut believes in the *Liebestod*, the death that connects, the Mother, Night.

War as the sweeping force for separation and loss, war as the analogue of the violence wreaked by parents on children, war as the fusion road, culminate in *Slaughterhouse-Five or The Children's Crusade*. Science fiction and realism coalesce, public and private horrors flow into each other in Vonnegut's vision of war as natural and impersonal as glaciers. In the Dresden fire-bombing of 1945, Vonnegut and other POWs spent the night in a meatlocker well beneath the earth and survived the fire storm that exploded above them. This real event is Vonnegut's most perfect symbol for the way many of his characters survive by burying themselves. Billy Pilgrim, the novel's hero, is a POW who has always been a POW: by World War II he is a casualty for the second time. His Great War began in childhood in the YMCA when his father "told him he could learn to swim by the method of sink or swim." His "father was going to throw Billy into the deep end of the pool and Billy was going to damn well swim. It was like an execution. Billy was numb as his father carried him from the shower room to the pool. His eyes were closed. When he opened his eyes, he was on the bottom of the pool. He lost consciousness, but the music went on. He dimly sensed that somebody was rescuing him. Billy resented that." Billy never gets angry at his father for throwing him in or taking him out. He is one of Vonnegut's many crucifieds who fight brutality by shutting it out of mind, burying themselves at the bottom of the pool, bottom of the earth, bottom of the universe. Billy's World War I happens so early that nothing that comes after is that much of a surprise. He does not fight back in war; he gets unstuck in time and space and lets his mind float free.

Vonnegut's orchestration of World War II with science fiction flights to Tralfamadore juggles outer space with spacing out, public catastrophe with personal anguish as it develops a working defensive system against pain in all its intensities. What barricade against a city's destruction? When the guards emerge from the meatlocker and see their incinerated city Vonnegut has them draw together "like a silent film of a barbershop quartet." Vonnegut writes them dumb. And dumbness is precisely his solution. Like Bokonon, he sends you to the ice death, tells you to go dead to the fact that the past is the destruction you have known, the present the violence you see, and the future the holocaust to come.

Spacing out is Vonnegut's answer to death, war, human glaciers. In Vonnegut's many descriptions of Pilgrim's trips to Tralfamadore, space-time travel is the ultimate withdrawal, the burial of suffering in meaninglessness. When you are looking at life from Tralfamadore, its wins and losses do not count. Tralfamadore provides not merely the vantage point *sub specie aeternitatis,* but the chance to see life in the context of intergalactic pointlessness. Billy Pilgrim space-times out to discover from one Tralfamadorian:

> On other days we have wars as horrible as any you've ever seen or read about. There isn't anything we can do about them so we simply don't look at them. We spend eternity looking at pleasant moments—like today at the zoo.

Later, Billy writes letters to newspaper editors, telling why he no longer weeps at the loss of anyone: "Now when I myself hear that somebody is dead, I simply shrug and say what the Tralfamadorians say about dead people, which is 'So it goes.' "

Vonnegut's Tralfamadorians, like his audience, cannot deal with death so they reduce its emotional charge, throw in the towel, and look the other way. When a little German girl asks Billy Pilgrim what he does in the war, he answers, "I don't know. I'm just trying to keep warm." "I know," says the little girl, "all the real soldiers are gone." In Vonnegut's world there is not one left who can fight against the destruction of war because everyone is already ground down by the disasters of peace, by cataclysmic lives that leave them too crushed to resist. Vonnegut's people see themselves as programmed for maximum disaster; their moments seem structured for the apocalypse because having been through it in childhood, they go through it again and again, irresistibly drawn to the bottom of the pool.

Vonnegut cares about how hard it is for the average man to feel alive and warm. Pilgrim, having been thrown into the deep end as a kid, grows up to fade away from caring deeply about his life—he stays in the shallows of emotion. He marries a 200-pound girl "no one in his right mind would look at" because his family thinks she is a good catch—her father has money. He space-times out to Tralfamadore where he is cozily caged with Mona Wildhack, a gorgeous porno movie star. On Earth he is a middle-aged optometrist, patiently fitting glasses on a mongoloid—like Vonnegut he is trying to make us idiots see better. What Vonnegut makes you see are the fantasies of the ordinary man replete in his dissatisfactions, his sense of having gotten little in the past and having nothing to look forward to.

The disaster people are programmed for in all Vonnegut's fiction is not death, but life. Against its power to wound, what defense? Passivity. Acceptance. Resignation. Denial. Looking the other way. Vonnegut celebrates the maneuvers of his audience caught in the marriage, the situation, the job it cannot change. His answer to the suffocation at the bottom of the YMCA pool, to the incineration of Dresden, is to can whatever pain they generate, to label it in clear, pure American colors. Vonnegut packages spaciness, resignation, denial as labels on pop graves. He writes as an epitaph for a friend shot amid the ruins of Dresden for taking a teapot, EVERYTHING WAS BEAUTIFUL AND NOTHING HURT. He invents a death rite where the healthy man and the dying man sit facing each other shouting "I loved everything I saw." And Vonnegut is full of tenderness for the benumbed, chilled man who kills his nerves so as not to kill himself.

Writers who are culture heroes speak to the audience's sense of the grinding joylessness of American days. Vonnegut gives up on people coalescing, loving, helping each other through; his humor is at the absurdity of hoping for uncut love. His many mental gymnastics are means of getting on without getting together. Other culture heroes also appeal to the audience's discontent. But they write about the attempt to carve out of tedium some area of satisfaction. Robert A. Heinlein's *Stranger in a Strange Land* is a parable about everyman's estrangement from pleasure, of how alien his attempt to find it makes him seem.

Heinlein's hero is Michael Valentine Smith, an ultimate isolate, who was abandoned in infancy on Mars, the sole survivor of an Earth mission. Raised by Martians, Smith is an alien on Mars, a stranger to Earth; he is another Billy Pilgrim, a total innocent befuddled by the universe. He is Heinlein's best vehicle for a comment on American ways. Heinlein's Mars is our Earth, our Earthling habits magnified. Martians are either bouncy, furry nestlings, full of life, or Old Ones who seem to be disembodied consciousnesses. A nestling chooses to discorporate into an Old One. He leaves his body to be eaten and perhaps "grokked" (spiritually encompassed) by his friends, and becomes a mind that ruminates forever, detached from any consideration for life. He never thinks about sex because there is none on the planet.

It is not Martians who choose death, but people. It is people who are brought up on devouring each other and come to see group cannibalism (war, genocide) as a form of togetherness, the fusion over and in destructiveness. People engraft death into their lives when they live without sex and kill their feelings. Michael Smith has been taught absolute cool on sexless Mars. He is an old-fashioned good baby, totally obedient to the Old Ones who turn him toward the totally cerebral life. The perfect passive, Smith knows how to go dead. When emotions rise (other people's, his own cannot) he curls up, slows his heartbeat, his respiration, and goes into suspended animation. When he reaches Earth as a young man and perceives the intensity of anger in people he asks, dismayed, "How can these human brothers suffer intense emotion without damage?" They can't! They can't! implies every line of Heinlein's novel, in which the world is packed with lonely, sexually hungry humans who have no control over their unhappiness. How to seize power for a revolution of feeling?

"The miracle of bipolar reproduction!" Sex, which Smith says is unique to Earth, is the means of gaining power over your own capacities for pleasure. Once Smith discovers it he worships it as the highest form of grokking, of conceptless knowledge, of incorporating with another person. He forms "nests" of people who join in collective sex, tied together through empathy and orgasm in groups where no one gets jealous, possessive or resentful. You can ride a sexual high to fusion, to connection with yourself and other people in Heinlein's vision of Earthling eroticism fused with Martian emotional control.

Group sex in Heinlein's novels is a means of finding your body, your pleasure, by losing your past, past personality, past inhibitions, past life. The synchronized group ethos that replaces individual marriages or exclusive affairs is a reflection of his sense that pleasure really lies only in the end of ego. People who cannot relinquish their need for exclusiveness leave the nests or are cast out, back to the world of jealousy and unhappiness. In this 1961 novel, Smith is stoned to death by a mob for immorality. But Heinlein tapped the audience's rising discontent with marriage, the unrest over close, emotionally draining relationships, the dissatisfaction with family life, the sense of powerlessness that were the coming themes of the decade. Sexual sharing is his means not only of gaining control over your body by gaining control over your life, but of breaking down the old personality that ails you in a group mind concentrating on pleasure. If in 1961 this vision of life as salvageable through sex was Heinlein's fantasy, it looms larger and larger as the national ideal.

Like Vonnegut and Heinlein, Arthur Clarke is preoccupied with erasing the yearning "I," the individual discontent. His *Childhood's End* opens with the kind of world Heinlein pleads for. Marriage exists, but only as a contractual arrangement for a limited time. Everyone recognizes the finite, ephemeral nature of affection and no one is bothered by it. This reasonable approach is the gift of Overlords, super-rational beings from a distant star who have taken over the management of earth so cleverly they've eliminated war, poverty, disease and even marriage unto death. The good life in this fifties book is filled with sports, entertainment and self-expression through consumption of luxury goods in a kind of grand suburban fantasy. But not everyone is happy. The world is still full of ambitions which cannot be fulfilled. Is there any way out of human discontent?

Clarke says that human evolution is moving toward a perfection point. In his theory of evolution, based on Einstein's theory of relativity and Teilhard de Chardin's mysticism, human stuff—personality, body, civilization—is moving toward ultimate conversion into pure energy. Though an impersonal biophysical process (no one is responsible!) a generation is produced that is no longer human. These last children require no emotional responses, have gone beyond feeling, are self-sufficient, self-contained, mute by choice and so uninterested in the physical world they do not even bother to open their eyes. They do not have egos. They are physically isolated, but interconnected because they plug into the same energy source: the Overmind, the massive molecular flow of the universe.

Clarke is obsessed in this novel with human limitations and possibilities. The Overlords, who represent the best possible achievements of reason, see themselves as the eternal servants of the Overmind, incapable of merging with it because of a genetic incapacity for intuitive, emotional, telepathic connection. Through them, Clarke expresses his sense of the helplessness of the most acute minds in the face of pure experience. The numb, unspeaking, blind children who finally dematerialize themselves and disintegrate the planet into the amorphous energy flow have fulfilled the highest possible destiny. Their destruction of everything is a radical metaphor for the belief shared by Vonnegut and Heinlein that to survive well you have to be an anonymous particle, an egoless, identityless part of the flow. The less human you are, the better chance you have to endure.

The appeal of Clarke's novel comes from its fusion of magnificent eeriness with the immediately recognizable situation of a generation of "children" at odds with the "good life" and determined to be invulnerable to the dominant culture. Aesop used the animal fable to tell some dark truths in a light way. Clarke uses science fiction to provide a vivid yet totally unthreatening portrait of America in the fifties and a prediction of the unrest that would erupt in the sixties. Does suburbia inevitably breed its own enemies? Is the overriding human need not for the rational life in the rational society but for wordless fusion with other people? Clarke spoke to the deepest emotional needs of the young audience. Timothy Leary in *The Politics of Ecstasy* would reinterpret those cravings in terms of a millenium of "precellular awareness" ushered in by LSD. Tripping,

> your nerve cells are aware—as Professor Einstein was aware—that all matter, all structure is pulsating energy; well there is a shattering moment in the deep psychedelic session when your body and the world around you dissolve into shimmering lattice-works of pulsating white waves, into silent subcellular worlds of shuttling energy. . . . Suddenly you realize that everything you thought of as reality or even as life itself—including your body—is just a dance of particles. You've climbed inside Einstein's formula, penetrated to the ultimate nature of matter, and you're pulsing in harmony with its primal, cosmic beat.

Orgiastic fragmentation is the leap beyond personality and emotion. Clarke and Leary tapped the growing sense of the abrasiveness between people, the fear that connection could never again be possible in terms of empathy, love or affection. Would human disconnection work better? Or self-subordination to a chemical process? Clarke's collaboration with Stanley Kubrick on *2001* welded Clarke's mysticism with Kubrick's machine mania into an imagistic assessment of life and death as mechanical processes. Dismantling a computer involves its regression down to its first memory bank; death involves switching off human functions and the return to the embryo. The film celebrates life as the infinite movement from one state of isolation to another—from encapsulation in spaceships and oxygen helmets to the primal encapsulation in the amniotic membrane Kubrick projects as large as the Earth. The secret of eternity is ceaseless and isolated transformation; wisdom is the recognition that in all your life, in all your future lives, you will be alone. You're born programmed for solitude and submission to the shattering power of the universe. This is one of the most popular statements of what kind of synchronization is possible for people. As the walking mud we all are in Vonnegut's *Cat's Cradle*, the replaceable lovers in Heinlein, or the nuclear bits we become in Clarke, what can bind us together is not ourselves, but our common anonymity, our reducibility to neural arrangements.

Can you endure and still be human? Can you get as far from pain as Mars without leaving Earth? Richard Brautigan has written of the healing force of ice in the blood. From *Trout Fishing in America* through his collection of stories, *Revenge of the Lawn,* to *In Watermelon Sugar,* his utopian novel, Brautigan has written of anger losing its heat. All Brautigan's characters are trout fisherman fishing for cool, freezing away every psychic ache, or looking for that cold hard alloy Brautigan calls "trout steel." "Imagine Pittsburgh," Brautigan asks, "made out of trout steel, the clear snow-filled river acting as foundry and heat." For Brautigan is the prophet of cities built out of ice rather than fire, of an America whose emblem would be no war-god eagle, but an elusive cold fish.

Brautigan's work is really one vision of people who successfully drown their feelings and lead underwater lives. His fishermen do not want to catch trout so much as they want to be like them. Brautigan's people can turn fury into a sweet mirage, can twist their private vendettas into a peaceful fantasy. "Revenge of the Lawn," the title story of his collection, is a deadpan story of vengeance. With no emotion a grandson tells of his gentle grandfather, a "minor . . . mystic" who prophesied the exact date World War I would start. But the very anticipation of violence drove him mad. For the rest of his life "he believed that he was six years old and it was a cloudy day about to rain and his mother was baking a chocolate cake." In Brautigan's world, this fantasy is war protest so effective it can replace the image of battle with a goody.

The grandmother, a flinty bootlegger, takes a lover who delights in destroying the lawn the grandfather created and loved. But, whereas the grandfather could not endure a fight and the grandson is stricken deadpan by everything, the *lawn* fights back. Becoming hard and malignant, it wreaks all kinds of havoc on the lover. Brautigan infuses so much violence into the lawn and so totally strips every passion from people that his message is clear. Only a lawn *could* fight back. No human could survive the rage he would feel if he let himself feel at all; no one could endure the outrage life engenders. Brautigan's people always submerge their feelings, always retreat from turmoil into a child-like innocence or a coldness so total that no passion, not even love, can intrude.

"Corporal" is a touching account of how people get to be so cool. A poor schoolboy during World War II yearns to be a general. In a paper drive his school organizes like a "military career," he scrounges for scrap after scrap of paper, hoping to bring in enough to spiral from private to general. But after an incredible effort, he finds all his work will make him no more than a corporal. (Only kids whose parents are rich enough to have cars and to know "where there were a lot of magazines" get to be officers.) Crushed and humiliated, he takes his "God-damn little stripes home in the absolute bottom of [his] pocket . . . and entered into the disenchanted paper shadows of America where failure is a bounced check or a bad report card or a letter ending a love affair and all the words that hurt people when they read them."

Suffering makes Brautigan people gentle and cold; humiliation turns them harder than trout steel and meek as fish. The grandfather and grandson feel so hurt they nullify themselves. Brautigan people fade away from competitive strife, from those wars for power and position that churn out losers ever more cruelly. And withdrawal and protection are their only answers to American aggression. "Revenge of the Lawn" is full of people taking shelter: a newsboy runs his paper route in an armored car, a child crawls into a hollow rock and pretends to live there.

What alternative to the isolated life underwater, underground, in a bullet-proof shell? Escape to a collective of isolates! Brautigan turns withdrawal into a strategic maneuver in a mental military show, a revolution of imagination. In "The Confederate General from Big Sur," Brautigan's rebels gather together in the woods for a gentle communal life. But in *In Watermelon Sugar* they try to reconstitute civilization by reconstituting people.

Euphoric, serene, *In Watermelon Sugar* is Brautigan's statement that life's possibilities really shrink to one emotional either/or: utopian iDEATH or the hellish state of inBOIL, either detachment and ego death or seething destructiveness. "In watermelon sugar the deed was done and done again as my life is done in watermelon sugar," says Brautigan's nameless iDEATHian narrator. iDEATH is a place where everything—houses, sensibilities, windowpanes—is made of watermelon sugar or trout steel, where human stuff is reconstituted from sweetness and coolness. There is no "I," no ego-striving, no marriage, no exclusive sexual passions, no professional envies, no ego games. No one has to do anything he really dislikes in this town that rises around a sacred trout hatchery, that can be reached over a bridge lit by watermelon sugar lanterns with the shape of a trout and the face of a beautiful child. Sweetness and light are everywhere. The innocence of fish and children is what peace demands. iDEATH requires not merely the death of the "I" in a communal mind, but the forgetting of the ego-mad human past, the abandonment of all the Pittsburghs built out of heat, sweat and fire.

Not far from the town of iDEATH are the FORGOTTEN WORKS with their mile-high ruins full of gadgets and books, the American skyscrapered world Brautigan would like to forget. This is the world of inBOIL, man in perpetual discontent. inBOIL is an iDEATHian misfit who moves into the ruins, brews whiskey from forgotten things and gets drunk on the past. He locks himself into old emotions of violence and egomania, the destructiveness and self-destructiveness Brautigan sees as the only emotions possible for the man who has emotions at all. One day inBOIL marches toward the sacred trout hatchery, screaming, "Are you afraid to find out what iDEATH really means?" It's not the erasure of the self in a group, not the freezing over of feeling, but blood-and-guts literal extinction. On the hatchery dance floor he cuts off his thumb and drops it into a tray filled with trout just barely hatched. His few followers obligingly follow him, cutting off their thumbs, noses, ears, until they bleed to death. Meanwhile, the only girl in iDEATH to get unattractively possessive of her lover commits suicide. Brautigan's "enemies" always have the good manners to kill themselves. iDEATHians mop the blood from the trout hatchery floor, cremate the corpses in their shacks in the FORGOTTEN WORKS and have a trout hatchery party.

In Brautigan's utopia you really can mop up human madness, wring it into a bucket and throw it out. When the iDEATHian narrator remembers his past, he wrings it free of emotion. He tells you of his boyhood in this anecdote of how tigers came in one morning and ate his parents:

> "Don't be afraid," one of the tigers said. "We're not going to hurt you. We don't hurt children. Just sit there where you are and we'll tell you a story."
>
> One of the tigers started eating my mother. He bit her arm off and started chewing on it. "What kind of story would you like to hear? I know a good story about a rabbit."
>
> "I don't want to hear a story," I said.
>
> "OK," the tiger said, and he took a bite out of my father. I sat there for a long time with the spoon in my hand, and then I put it down.
>
> "Those were my folks," I said finally.
>
> "We're sorry," one of the tigers said. "We really are."
>
> "Yeah," the other tiger said. "We wouldn't do this if we didn't have to, if we weren't absolutely forced to. But this is the only way we can keep alive."
>
> "We're just like you," the other tiger said. "We speak the same language you do, we think the same thoughts, but we're tigers."

"You could help me with my arithmetic," I said.

"What's that?"

"My arithmetic."

"Yeah."

"What do you want to know?" one of the tigers said.

"What's nine times nine?"

"Eighty-one," a tiger said.

"What's eight times eight?"

"Fifty-six," a tiger said.

I asked them half a dozen other questions: six times six, seven times four, etc. I was having a lot of trouble with arithmetic. Finally the tigers got bored with my questions and told me to go away.

"OK," I said. "I'll go outside." . . .

After about an hour or so the tigers came outside and stretched and yawned.

"It's a nice day," one of the tigers said.

"Yeah," the other tiger said. "Beautiful."

"We're awfully sorry we had to kill your parents and eat them. Please try to understand. We tigers are not evil. This is just a thing we have to do."

"All right," I said. "And thanks for helping me with my arithmetic."

"Think nothing of it."

The tigers left.

I went over to iDEATH. . . .

Brautigan's will to crack anger into unrecognizable forms is evident in this episode, which reflects what the narrator feels destroyed his parents, how they treated him, and how he would like to retaliate. The tigers are embodiments of the rapacious hungers that devoured his parents' lives and made them tigers to him. Part of what gets chewed up is the narrator's boyhood. In this dream-like anecdote, he is using the ferocious, super-human beasts to do to his parents what they have done to him. The humans who are devoured are as child-like and passive before the towering, clever animals as a little boy would be toward the parents who seem totally in control of his life. This vision of being eaten alive in childhood by your parents, by the best Authorities! is Brautigan's vision of what life outside iDEATH is like, of what you would see if you remembered the forgotten works of your own past. The memory is permissible—even possible—for the narrator only in alien imagery and with the beautiful idea that tiger-people are extinct.

Splitting the "I" from life, welding an "i" to DEATH means not merely subordinating the individual consciousness, or ending ego-striving in a group tuned in to peaceful low-key vibes, but separating yourself from your own, anger-ridden past. What Brautigan wants are those easy feelings that flow only from the chosen present, from a world without enforced relations, a world without associations, without real memory. Brautigan always writes of the tiger-bitten, the helpless onlooker, who alchemizes himself into a trout-person, who lives with steely passions and diluted hopes. Bruatigan sees cutting out your heart as the only way to endure, the most beautiful way to protest the fact that life can be an endless down, the perpetual encounter with cruelty in others and yourself. Brautigan's asset as a writer is his verbal wildness, his simplicity, the passive force of his people who have gone beyond winning, losing, loving or hating.

"You would be a hero in Berkeley," Vida, a gorgeous girl, tells a trout-man in *The Abortion: A Historical Romance, 1966*. And Brautigan became a hero there and elsewhere because his lyrical novels resonate with the depression and the hopes of an audience that feels ground down, gnawed by tigerish families, pushed into the tiger-world, and hurt in its intense loves. Like the librarian hero of *The Abortion* who gathers books that lonely, tormented, ordinary people write to themselves and who places them nicely on a shelf in the America Forever library, the best any of Brautigan's people hope for is to shelve themselves away from competitiveness, from the fight for status and money, for self-assertion and success that turns people into tigers.

The view from Mars or Tralfamadore, from the Overlord's star or from iDEATH, is a vision of conceptual and emotional alternatives to powerlessness and insecurity, to the fearfulness of our connections with other people. Vonnegut, Clarke, Heinlein and Brautigan are all obsessed with ideas of fusion that they express in weird, alien images, but that are recognizable as pleas for the obliteration of the hierarchies of status, money, intelligence and aspiration they feel divide people from each other.

Are you lucky mud, divine molecules, unique sexual synapses, part of a heavenly-dead group mind? Can you give up the old American "I," the self-assertive urge, the adversary-dominated competitive life in the recognition of shared hardship, the embrace that says you are your enemy and he is you? Clarke, the total anti-materialist, writes of every product of intelligence disintegrating, the absolute conversion of art and consumer goods, of the gifted and the drone, to pure energy. Brautigan's people do not play the game, so they cannot possibly lose. Heinlein's "sex nests" are the happiest expressions of the hope that interchangeability is the answer to possessiveness, jealousy, the cynical joylessness of the rest of the world. All three writers reflect the hope for the disappearance of differences through physical and psychic coalescence. And they mirror how profoundly people have given up on trying to control anything but their bodies.

A culture creates its heroes even as its heroes create novels. Vonnegut is the hero of culture heroes because he celebrates standard American vulnerability. Ordinariness, sameness, the sheer mindlessness of heroes who are poor slobs like the rest of us is one pole of Vonnegut's fiction. Fragmentation in outer space, the easy breakup into mud, is the other. Both meet in Vonnegut's belief in taking the step beyond personality. Vonnegut sees wholeness and satisfaction as inconceivable. His ideas of fusion through anonymity reflect a male flight from frustration, the greatest possible withdrawal from taking charge. They reflect a widespread will to get beyond the confines of individual responsibility and out of the single, beleaguered self. Vonnegut's ability to be that ravaged figure has made him one of the most beloved writers in America. His novels arouse not only admiration, but protectiveness: his least works are praised along with his best. He offers an unthreatening, instantly recognizable portrait of man ground under the heel of American expectations. Vonnegut reassures. He knows what most men achieve is not satisfaction—the loving wife, the status job, the son to be proud of—but resignation and perseverance in a numbing and draining condition.

Vonnegut's message is that beyond bleakness and muddiness of mind is something worse: knowledge of how bad your life is, the pain beyond endurance, the delusion that there is something better. Men who believe in something, who try to take charge of their lives, become in Vonnegut's fiction, the true buffoons, the nuts, the Dr. Frankensteins, the moralists. Like the noble high school teacher in *Slaughterhouse-Five* who speaks against hate to a Nazi propagandist but who gets shot taking a teapot amid the ruins of Dresden, the good not only do not survive, they are killed by others with high moral standards. Vonnegut seems to believe that only fools and crazies let themselves care, or let themselves distinguish between one kind of madness and another. What Vonnegut prescribes is not the freedom to be yourself, but the freedom to be nothing.

Vonnegut's antidote is dullness. He insists on the right to throw down the burden of individuality, the troublesome spark of love for a value, a woman, a cause. His true hero is the man slightly cracked by his own frustrations, but holding himself together with his triviality—his aluminum siding, his humdrum marriage, his empty job, the boredom that keeps him from thinking or feeling.

Vonnegut is the champion of an American audience that would like to space out of its heart and mind so as not to have to hurt. He writes about men of today, ordinary astronauts. Neil Armstrong in trouble, as Norman Mailer said, "talked computerese." Buzz Aldrin, asked about stepping out onto the moon, replied, "The touchdown itself will be the ultimate test on the landing gear." But not on the man, the mind. Vonnegut is the writer who has become our star moon man, who shows the way anyone can dump his mind and heart and space out to an orbit where, whatever happens, it really seems A-OK. Of course, nothing is OK unless you can believe that the answer to what bothers you is not caring about what bothers you.

III

ANGRIES: S-M AS A LITERARY STYLE

Murder only deprives the victim of his first life; a means must be found of depriving him of his second. . . .

de Sade

You have conquered countless fierce and barbarous nations, and infinite lands rich in every kind of treasure. But when you conquered these you conquered what had a nature to be conquered. There is no strength so great that it cannot be weakened by force of arms. To conquer the mind, to subdue one's own anger, to temper victory, and to not only raise up a fallen foe of the first nobility, talent and virtue, but also to enlarge his former dignity—he who

> does this, I do not compare with the most illustrious men, I judge him as most like to God.
>
> Cicero,
> "The Pardon of Marcellus"

Anger so bristles out of every American city that it seems to be the one irresistible emotion. Some fiction strips every embrace, every encounter to a power struggle with a thoroughness that reflects the audience's fear that there is nothing left but combativeness. In this fiction the street and the dining room are equally full of war, the hater and the hated equally guests at the same American feed on aggressiveness and fear. S-M fiction is perversely holistic. It simplifies, organizes all experience into a power struggle and pushes to the limit the value placed on performance and achievement. It mirrors the urgency with which some people feel pride depends on proving yourself no victim by becoming a fury.

The popular equation of economic and sexual repression and the blurring of distinctions between social and personal revolution have made S-M fiction the ready vehicle for describing the outcast, the murderer, the junkie. It permits the novelist to deny his own anger and to write from the position of the observer, or the victim who fears colossal political malice or dreads someone else's anger. But the increasing experience and tolerance of violence have made it possible for some to admit that what they claimed to fear, they would like to do. Yesterday's paranoiac is today's closet murderer.

William Burroughs, for years the underground's loudest

voice against social, political and sexual repression, wrote in his novel *Wild Boys* of packs of homosexual guerillas dedicated to destroying people: "The family unit and its cancerous expansion into tribes, nations we will eradicate at its vegetable roots." His "boys" race into the "chintzy middle class living room throwing gasoline on the nice young couple watching TV. One wild boy lit a match [his] face young pure pitiless as the cleansing fire" that envelops the young couple. "He lit a [cigarette] with the same match, sucked the smoke in and smiled. He was listening to the screams." Asked in a *Rolling Stone* interview, "Is the book a projection?" Burroughs replied, "Yes. It's all simply a personal projection. A prediction? I hope so. Would I consider events similar to the *Wild Boys* scenario desirable? Yes, desirable to me."

Burroughs' novels have worked toward this smug, vindictive fury. Although he has been called a prophet of political doom and a satirist of the American scene, Burroughs has in fact never written against institutions so much as about emotion, never against ideas so much as about the experience of degradation or fear. His political outrage has the meaning of expletives strung together with great intensity. What it communicates is not content but energy, that diffuse, furious revolt that gives his work the edgy power of acid rock. His explosive style is what makes him an exciting writer. Burroughs' cut-up or fold-in method (take a page or tape, cut it into pieces, and paste or splice together in any order except the original one) is not only designed to destroy any coherent statement, but to dissolve the line between personality, society and politics in a jolting music. This very confusion has served to conceal Burroughs the writer and the hater amid his many indistinguishable characters, those anonymous voices choking with nausea at being alive. For Burrough's final subject is not the particular psyche, nor the social scene, but quaking nerves.

Burroughs' novels are the experience of his orgiastic hate. For him every emotion is a disease, every urge has the driving, manic force of compulsion. Passion is the work of demon bacteria, "virus powers" that infect with love or a craving for heroin, cruelty or power. Men are no more than hosts for personality parasites, victims that virus-gods use for food. And the sex parasite is the most voracious. For Burroughs always writes of those lonely, homosexual men who are doomed to be tempted into sex by the likes of Johnny Yen, "the boy-girl god of sexual frustration from the terminal sewers of Venus." His characters tell of lust rolling into degradation and torture, of beautiful boys metamorphosing into slimy green newts, of the thousand monsters that ripple out of frustration. And as Burroughs politicized heroin addiction in *Naked Lunch* as the mold of all addictions—of governments to power, of men to cruelty—so in his later novels sex becomes the model for every kind of conflict and degradation. Living only means being wrecked by desire, being violated by men who are not even people but like those mutants whose "penis had absorbed the body," leaving only "vestigial arms and legs."

Burroughs' spaced-out men are always changing and always the same. They are really one "vocal apparatus" with attached "metabolic appliances," one body plugged like a soft machine into an erotic energy source. They are not people at all but abstractions of human drives. They move among "flesh gardens" where humanoid plants ejaculate, remember "white sheets dripping nova," or again and again recall hanged men ejaculating in death until orgasm itself seems like an impersonal, lethal energy. But no matter how far they recede into abstraction, they can never escape the sense that every eruption of life is a force for destruction. Burroughs' titles—*Nova Express, The Ticket That Exploded*—tell nearly all: life is a fulminating train speeding toward its final, explosive stop. Your ticket to ride is your body, the raging host that makes you ride whether you will or no. How to stop the train? How to escape the "orgasm death"? How to avoid Johnny Yen who promises "love love love in slop buckets" but is nothing but trouble? How to avoid self-loathing and fury?

In *Naked Lunch, Soft Machine* and *Ticket That Exploded,* Burroughs wrote of people consumed by their own passions, of lovers spliced together in hate, their intestines locked in a parasitic bond. But in *Nova Express* and *Wild Boys* he evolved a way of dealing with every agony. In these novels, a man can escape his hate by going out of his mind. Burroughs turns emotion into a psychic event that releases feeling in a brilliant, removed image: over and over he describes astral bodies caught in concentrated sexual agony. Mr. Bradly and Mr. Martin are stars, locked together by gravitational forces. Mr. Bradly, a small dense blue star, pulls fuel from Mr. Martin, a large red one, and grows brighter while Mr. Martin is diminished. "First it's symbiosis, then parasitism," comments Burroughs on love. In *Nova Express* wars of sex and politics are waged in outer space. In *Soft Machine* they occur in the ancient past: a Mayan peasant is so bitterly used by a high priest that the tyranny of office reverberates with sexual cruelty. But the intercourse of stars or the degradation of a Mayan are more than parables for love or tyranny. Events like these reflect the process by which Burroughs' characters disown what happens to them, where it happens, and what they feel. In the remote recesses of hallucination they symbolize their rage away, move out into space or back in time, but always away from their own fury.

Alienation and spaciness have been Burroughs' smokescreens, concealing the pure intensity of his misanthropy. Even so, hatred and self-hatred surface in his many images of homosexual lovers who know that "murder is never out of his eyes when he looks at me" and "murder is never out of my eyes when I look at him." In *Wild Boys* Burroughs himself emerges apart from "Venusian" hates, no longer the victim but the master of those "virus-gods" that infect people with the love-and-need disease. And he is unmistakable as the Mayan super-priest, the "Incomparable Yellow Serpent" who is a demon artist. He controls everyone by "singing the pictures" that fill their minds and pattern their actions. He "shifts from AC to DC as a thin siren wail breaks from his lips. Pictures crash and leap from his eyes blasting everyone to smoldering fragments. DEATH, DEATH, DEATH. When comes such another singer?" demands Burroughs.

The images that spring from the pages of Burroughs novels are clearly meant to kill. Burroughs has always written of life as able to be controlled by self-images, by "sets" of behavior programmed or taped into consciousness. But he has usually not written from the point of view of the filmmaker. *Wild Boys,* however, is clearly meant to be a movie of the wild mind of Burroughs, "Billy B. St. Louis Encephalitis," who records the "1920 St. Louis boyhood of Audrey" whom everyone humiliated because he "looked like a homosexual sheep-killing dog," and who went to "kindergartens like mental homes." In fantasy films of the future, wild boys vindicate Audrey by erasing the word "mother" from the blackboard. More than cinematic in style, the novel works toward a statement of life as movie, of people as moviegoers—the spectators of their own lives—and of both life and film as products of Maya, the force of illusion that shapes all visible forms. "Nothing is true, everything is permitted" writes Burroughs as he kills off all the straights.

Burroughs' fantasies are true representations of his will. What he wants are "wild boys," the sudden coming of "a whole generation . . . that felt neither pleasure or pain." Wild boys live in an emotional nowhere with no "emotions oxygen." They have no memory because they have no past—some of them are born through a process of replication in which they spring full grown from another man, vibrate into life, and immediately begin having anal intercourse with their creator without desire or loathing. (This is Buddhism à la Burroughs.) What explodes from the wild boys' iciness is Burroughs' passion for purging all the old self-images memory harbors, all the hates that fester in the mind. In Burroughs' filmland, memory and emotion can be exploded away. When one man begins to remember "the pawn shops, the cheap rooming houses, the chili parlors," he detonates a "film grenade" and "explodes the set." The "boys" end the novel by wishing the world dead (the ultimate in social protest!) and watching the "screen explode in moon craters and boiling silver spots. They see "dim jerky stars blowing away across the empty sky." Having turned us all to ashes, "wild boys smile."

Burroughs uses fantasy and hallucination as devices for detaching himself from rage and concealing it from you. But despite the wildness of his imagination, the technological language, the machine people who are supposed to have no emotion, fury persistently breaks through. *Wild Boys* is an admission of his hate where denial almost gives way to acknowledgment so that his new defense is the non-defense of letting it all hang out.

Hubert Selby is a surrealist of the streets whose novels have the immediacy of primitive art. Selby's power comes from his ability to manipulate hate, embrace it as a fact, and love it because it is there. He prefaced *The Room*, his portrait of a festering mind, with the words, "This book is dedicated, with love, to those who remain nameless and know." From human dregs, from the unremittingly tormented, Selby extracts the very odor of rage, the essence of that free-floating anger that lies like a pall over all of us. In *Last Exit to Brooklyn* he wrote of the back alleys, the bars, the empty lots where petty crooks and local gangsters break out of the endless boredom of their lives into violence. Selby's Brooklyn is not a place so much as a nightmare where manhood can be won only through one man's torment of another. The relentless pursuit of machismo through all the ways of cruelty, the fear of failure and worthlessness that drives men into deeper and deeper vileness are Selby's preoccupations. He is a clinician of male violence, dissecting straight to the center of sexual chaos and cruelty.

As Harry, the hero of "Strike," rolls over on Mary, his wife, "hitting her on the head with his elbow, wanting to pile drive his cock into her, his anger and hatred started him lunging and lunging until he finally was all the way in—Mary wincing slightly then sighing—Harry shoved and pounded as hard as he could, wanting to drive the fucking thing out of the top of her head; wishing he could put on a rubber dipped in iron filings or ground glass and rip her guts out—Mary wrapping her legs around him, rolling from side to side with excitement." He rolls off, his "stomach crawling with nausea; his disgust seeming to wrap itself around him as a snake slowly, methodically and painfully squeezing the life from him, but each time it reached the point where just the slightest more pressure would bring an end to everything: life, misery, pain, it stopped tightening, retained the pressure and Harry just hung there his body alive with pain, his mind sick with disgust." This is one of Selby's rare heterosexual love scenes.

Most of Selby's characters hate women so totally that they do not want to get close to them even to destroy them. His most brilliant sequences deal with the affairs of "queens" and "johns," with those relations between men in which one exults in sexually terrorizing the other. One of his best achievements is his portrait of Georgette, a "hip queer," whose life "spun centrifugally around stimulants, opiates, johns." Out of a vertigo of bennies, gin and sexual hunger, Georgette loves Vinnie who is really "rough trade." She spirals higher and higher on her hope that she is the most interesting woman in his life, quivering over his final willingness to let her blow him, only to find that what she tastes is not his "love," but the unmistakable leavings of his anal intercourse with someone else. Welling from her mouth like recognition is her nausea at the humiliation of her life and the terrible, pathetic insistence that what she tasted "wasn't . . . it wasn't . . . shit." But in Selby's world it always is.

Where *Last Exit to Brooklyn* focuses on the social scene of violence, from brawls to sexual infighting, *The Room* is the story of one mind riveted on fury. Sadism is the only means of survival for Selby's nameless hero, who festers in prison as he awaits trial for an unnamed crime. His mind is filled with nothing but craving for revenge on the teachers, judges, cops who have put him down. Only in maiming another living thing has he ever felt whole or alive. But only over two dogs has he ever achieved that power: "What a supreme joy . . . to watch them crawl on all fours over gravel and broken glass and then when their palms were hardened so they could plod along as rapidly as possible, the callouses were peeled off and they had to start all over again. . . ."

This is Selby's vision of a culture's bedrock psyche, a portrait of an American mind gone the limit in its acceptance of cruelty as life's only fixed principle. Selby sees every act as for degradation, sees the basic bond between people as pain, whether inflicted or felt. If he does not gloat over the cruelty he describes, Selby nevertheless sees nothing else, nothing but the terror of those dismal, festering characters who spring so fully formed in their vileness from his imagination. He does write of them with love, with an energy and purity of style that is absolute in its insistence on your glimmer of recognition and assent. Selby aims at the audience's ever-ready sense of inadequacy and humiliation. He hits the mark because reading him is like being mugged.

The one surreal, the other primitive as fact, Burroughs and Selby write of hate as open fire, finally immersing themselves in fury as an experience to love. But there are writers who are into hate, but out of love with it, who use a variety of devices to reduce their anger. Donald Barthelme and Truman Capote deal with depression, with anger muted to pessimism and discontent. Rage, in their work, takes the form of despair over the possibilities for life.

Elegant, spacy, more than slightly modish, Donald Barthelme has such disdain for life that he estheticizes even his depression. His themes in his richest work, *City Life*, are absolutely rarefied: the exhaustion of creative power, the disintegration of individual consciousness, the artist's alienation from the ethos that produced a Tolstoy. Barthelme's best works are those with the strongest emotional charge. But his strongest emotion is anger. Is Barthelme's anger the writer's persistent frustraton with the bits and pieces of modern, fragmented personality? *City Life* is about, among other things, an artist's affliction and cure.

"The Glass Mountain" is a parable about an artist caught between an exhausting concrete labor, his own frustration, and the fury of his audience. The story parodies a fairy tale, partly retold in it, about a youth who must reach the top of a glass mountain to free a beautiful enchanted princess. But Barthelme's hero is a New Yorker painfully climbing a glass mountain on the corner of Thirteenth Street and Eighth Avenue with the aid of two plumber's friends. He wants to free an enchanted symbol at the top.

The climber inches along, aware that the mountain is already ringed by corpses of people who have made the attempt and failed. The streets around them are studded with dog shit and crowded with his audience—passers-by who shout obscenities at the climber, wonder who will get his apartment, and make such comments as "Won't he make a splash when he falls now?" He realizes he will never free the symbol through his labor; only a literal flight of imagination could do it. He borrows a method from the fairy tale he parodies: he seizes the legs of an eagle and is carried above the mountain until at the proper point, he cuts off the eagle's legs and lands upright on the mountaintop. And he even finds the enchanted symbol. But when he touches it, it changes into "only a beautiful princess." He flings her "headfirst down the mountain" to the cursing crowd who "can be relied upon to deal with her." Barthelme's artist deals with life exactly like the crowd below. At the heights and depths there is nothing but hatred for other people.

In "Brain Damage" Barthelme suggests an alternative to the vulgar and rarefied forms of rage: endless discontent, imaginative ruminations without belief in imagination's power. Disbelief in art, like pessimism about life, has a particular force for Barthelme. Meaninglessness is his answer to the rage of the glass-mountain artist who is always sliding into the fact of his own hate, who cannot find images anywhere that are removed enough from the intimacy even fairy-tale princesses suggest. In droning emptiness and trivial activity Barthelme finds a cure for brain damage (thought and feeling): "The elevator girls were standing very close together. One girl put a candy bar into another girl's mouth, and the other girl put a hamburger into another girl's mouth." Just as concentrating on a candy bar frees the elevator girl from an awareness of highs and lows, so the Winston ad, sung by a passing girl, saves another Barthelme character from falling apart, falling down into his wrath. Anything that produces enough boredom to blot out feeling and ward off the emotions that "damage" the mind stabilizes the collapsing personalities Barthelme loves. Feeling looms large only in questions between episodes, when "To what end? What recourse?" appear in heavy type. Barthelme's episodes never do more with these questions than show how to avoid either asking or answering them.

Barthelme's use of emptiness and boredom is his argument against the tense fury of the glass-mountain artist. His own excursions into meaninglessness and contemporary deadness are really an alternative to open fury. If there is nothing that signifies anything, then having or not having are alike. No one has to struggle with his desires, or experience the repressed, volcanic rage of the artist who cuts the feet off his own inspirational eagle and has nothing left to stand on.

The most chilling thing in *In Cold Blood* is Truman Capote's contempt for the Clutters, who were murdered by Perry Smith and Dick Hickock. Subtle, pervasive, Capote's judgment on the Clutters begins with his opening description of Holcomb, which stands "on the high wheat plains of western Kansas, a lonesome area that other Kansans call 'out there' . . . in the earliest hours of that morning in November . . . certain foreign sounds impinged on the normal nightly Holcomb noises—on the keening hysteria of coyotes, the dry scrape of tumbleweed, the racing, receding wail of locomotive whistles." These "normal" Holcomb sounds reverberate with Capote's comment on the plains people who are adrift between absolute loneliness and "keening hysteria," perpetually alone in an emotional "out there." Into this atmosphere of desperate barrenness, Capote infuses not merely the expectation of death, but the fact that it already exists, not as a violent event, but as a way of life.

It is living in cold blood, not merely murdering in it that fascinates Capote, that shapes his description of the Clutters into something less than kind. The father, Herb Clutter, such a pillar of the community, so determinedly getting life under control, so able to keep his emotions in check—the daughter, Nancy, pie-baking, girl-scouting, helping with a niceness and intensity too great to be believed, the mother, Bonnie—poor Bonnie—who could not keep up the pace of all this assiduous goodness and broke down—this assortment of characters ap-appears almost ruthlessly empty. Clutter lets his wife go from her bedroom to a mental hospital without trying to comfort her or bring her out of her collapse. Capote sees him as never stopping his do-gooding long enough to think about anything. The Clutters are as much an abstraction for him as they are for Perry who murders them, probably as symbols for everyone who has ever turned him down.

Against the Clutters' stark emptiness, Capote places his brilliant, complex, intensely sympathetic portrait of Perry, a half-breed drifter, a loser raised in orphanages where he was often beaten, where, at the age of seven, he first dreamed a dream so rich in symbolic connections, so baroque in its imagery, so beautifully violent that it could not help but appeal to an author whose books had hitherto been so lush. After being battered by a nun for wetting his bed, Perry dreamed that "the parrot appeared, arrived while he slept, a bird taller than Jesus, yellow like a sunflower, a warrior-angel who blinded the nuns with his beak, fed upon their eyes, slaughtered them as they pleaded for mercy, then so gently lifted him, enfolded him, winged him away to paradise." The "holy spirit" for Perry was violent revenge, a child's dream of a force great enough to overwhelm the forces that overwhelmed him. But certainly this spirit of vindication came to Perry Smith in reality, in the form of the writer who would have so much compassion for him, who would immortalize his suffering and his act as he stood facing Herb Clutter, knife in hand.

"Who is lonelier, the hawk or the worm?" Capote raised the question in *Other Voices, Other Rooms,* and it must have obsessed him throughout *In Cold Blood* as he turned the collision between Smith and Clutter into an encounter between two isolates. Clutter stands in his self-control as Perry, conscious only of the pain in his own leg, murders him so unaware of his act that he must infer it is he who cut Clutter's throat from the fact that he holds a dripping knife. But the psychopath's power to blot his murder out of mind fascinates Capote no more than Clutter's whole life of emotional repression.

The connection in murder between Clutter and Smith is between two emotionally dead men, the "hawk" and the "worm" who are equally locked in their own isolation. The one murdering, the other dying, they are only Capote's most dramatic symbols for human "interaction," for the unbearable connection and disconnection between people. Capote can treat this relation like farce in his concluding chapter, "The Corner," where a bunch of murderers become good friends on death row. There are York and Latham who look like "two cheerful, milkfed athletes" and who reply when asked why they killed twelve people, "with a self-congratulatory grin, 'We hate the world.' . . . 'It's a rotten world,' Latham said. 'There's no answer to it but meanness. That's all anybody understands—meanness. Burn down the man's barn—he'll understand that. Poison his dog. Kill him.' Ronnie said Latham was 'one hundred percent correct,' adding, 'Anyway, anybody you kill, you're doing them a favor.' " Or there is the conversation between Hickock and Andrews, the Nicest Boy in Wolcott, who one day killed his parents. Talking of Perry's starving himself on death row, Andrews says, " 'Well, it sure strikes me a hard way to do it. Starving yourself. Because sooner or later we'll all get out of here. Either walk out—or be carried out in a coffin. Myself I don't care whether I walk or get carried. It's all the same in the end.' . . . 'The trouble with you, Andy,' " Hickock replies, " 'you've got no respect for human life. Including your own.' "

The trouble with Capote is that he sees everyone caught in the corner, fixed between one kind of death and another. The murder of the Clutters forces the people of Holcomb to question how they live—even the FBI man out to get the killers is really investigating himself. As he watches a bonfire of the Clutters' effects the FBI man asks: "How was it possible that such effort, such plain virtue could overnight be reduced to this—smoke thinning as it rose and was received by the big, annihilating sky?" Between the "annihilating sky" and the "scraping tumbleweed" is Capote's vision of the American soul living, dying and murdering in cold blood. Capote's despair over the ruined Perry Smith is the index of his distance from the Clutters. But his insistence that cold hate and cold life are the American experience is probably his wish more than his fear. Is it better to deep-freeze fury than to feel it, or even see it? Setting out to be the reporter of a real, evil act, did

Capote encounter only his old nightmarish fantasy, the anguish of Joel, the boy in *Other Voices, Other Rooms* who is in perpetual danger of being unhinged by his own loneliness, who, "reading a volume of Scottish legends finds one which concerned a man who compounded a magic potion unwisely enabling him to read the minds of other men and see deep into their souls. The evil he saw and the shock of it, turned his eyes into open sores."

"Only connect," wrote Forster. But for Barthelme and Capote connection means connection to the evil, the anger or the death they see at the center of every human being. Barthelme treats fury as a problem of art. But the boredom and emptiness his characters use to keep themselves from caring about anything at all are more than an esthetic device. In the fiction of both Barthelme and Capote, knowing or caring about people is invariably the experience of evil and suffering. Barthelme's estheticism and Capote's journalistic "realism" serve as devices for distancing both the writer and the reader from the violent core of the work. What grips Capote in the world is finally what justifies the depression, the quiet frustration that runs through all his serious novels. Reality is where Capote's fantasies happen.

The Dogs Bark, Capote's collection of portraits, interviews, and travel memoirs, underlines what makes the world irresistible to him. Places are, for Capote, most vivid as stage sets for the perpetual combats in his mind. In Hollywood on a Sunday in 1946, Capote says, his "shadow" is "moving down the stark white street . . . like the one living element in a Chirico," where every house or "crack in the wall" can "strike an ugly note prophesying doom." New Orleans streets, he says, "have long lonesome perspectives; in empty hours . . . things innocent ordinarily . . . acquire qualities of violence." "Acquire"? How? Streets are magnets for Capote's venomous fantasies.

Capote infuses his own violence into the quietest, emptiest places; he brings it to his portraits and interviews with public people, particularly those men who have been among the greatest symbols of virility. Humphrey Bogart is neatly struck down in less than two pages for his stupidity. "A most moral—by a bit exaggerating you might say 'prim'—man, Bogart employed 'professional' as a platinum medal to be distributed among persons whose behavior he sanctioned; 'bum,' the reverse of an accolade, conveyed, when spoken by him, almost scarifying displeasure. 'A guy who doesn't leave his wife and kids provided for, he's a bum,' " says Bogart, even of his own father. "Bums, too," explains Capote, "were guys who cheated on their wives, cheated on their taxes, and all whiners, gossipists, most politicians, most writers, women who Drank, women who were scornful of men who Drank; but the bum true-blue was any fellow who shirked his job, was not, in meticulous style, a 'pro' in his work." Having nailed Bogart for having the moral sharpness of a meat cleaver, Capote gets Marlon Brando for his confusion. Interviewing Brando in Kyoto during the filming of *Sayonara*, he makes him seem ridiculous for his pursuit of philosophy, his professed ambivalence over his success, his perpetual appearance of discontent. Noting—as any interviewer would—the gap between the piles of books on Zen and the lacquered trays heaped with food brought into Brando's opulent hotel room, Capote reduces most of what Brando says to merely the pretensions of a "boy on the candy pile."

Capote's mannered, ad hominem attacks so absorb his energies that he will not acknowledge that the enormous success of Bogart and Brando was rooted in precisely what he mocks them for. Brando and Bogart were not merely actors each playing "forever the same part." They were culture heroes whose differences suggest the gap between Americans' feelings about World War II and the Korean War, and the decline in our own security about ourselves. Capote uses his wit to blast all symbols of self-assurance—Clutter, the man with self-control, Bogart, Brando all get slashed in Capote's vendetta.

What does Capote admire? Jean Cocteau's ability to be "vastly imaginative but vivaciously insincere." Capote quotes Gide's journal describing a meeting with Cocteau in August 1914:

> His conversation shocked me like a luxury article displayed in a period of famine and mourning. He is dressed almost like a soldier

> and the fillip of the present events has made him look healthier. When speaking of the slaughter of Mulhouse, he uses amusing adjectives and mimicry; he imitates the bugle call and the whistling of the shrapnel. Then changing subjects . . . he claims to be sad . . . then he talks of the lady at the Red Cross who shouted on the stairway, "I was promised fifty wounded men for this morning: I want my fifty wounded men."

Capote has Cocteau's unerring instinct for the ghoulishness of everybody—even himself. But he thinks it's funny. He passes it off as coy wit in "Self-Portrait" where he asks himself, "Have you ever wanted to kill anybody?" "No? Cross your heart? As for me, if desire had ever been transferred into action, I'd be right up there with Jack the Ripper." Some fun!

Capote's glossy sniping is hate as chic, sadism as an Olympian attitude of Hollywood and New York. Does an establishment that institutionalizes cold contempt adopt the rip-him-off-before-he-rips-you-off ethic, or spawn an S-M underground, an East Village statement that the biggest success is the biggest sadist? Whether it does or not, the S-M spirit is aloft in the counterculture, flying on paranoia and speed. There are no major new writers in the subterranean hate-circus, but Hunter Thompson's reportage appeals to an audience willing to swallow S-M as social history.

Thompson comes on as a hip rebel in *Fear and Loathing in Las Vegas,* concerned about all the "right" things. But he says the failure of the Youth Movement was sealed when Allen Ginsberg and Ken Kesey could not persuade the Hell's Angels to join the Berkeley radicals. What the movement needed was presumably more kinkiness. What Thompson obviously needs is the combat, the invective, the insult. What he likes is not bringing down the establishment, but the chance it provides for a showy rip-off. The kick in *Las Vegas* is to drive to a posh hotel in an Expense Account White Cadillac, rent a room, and stock up on cocaine, red blotter acid, ups, downs, and other goodies—all at "their" expense—and to drug up in the middle of a convention of narcos. And there is the joy of outraging a maid who has a "face like Mickey Rooney," who opens a closet and stares down at Thompson's attorney" who is, "kneeling, stark naked, in the closet, vomiting into his shoes, thinking he was actually in the bathroom."

Vomit along with meeee! To be in you have to join the togetherness that wells up over the shoe and bowl. One of Thompson's friendliest conversations with his attorney is about how the attorney ate a whole Jimson weed in twenty minutes. "What happened?" demands Thompson. "I vomited most of it right back up. But even so I went blind for three days. I was such a mess that they had to haul me back . . . in a wheelbarrow . . . they said I was trying to talk, but I sounded like a raccoon." "Fantastic," says Thompson. But "I could barely hear him. I was so wired that my hands were clawing uncontrollably at the bedspread . . . I could feel my eyeballs swelling about to pop out of the sockets. 'Finish the fucking story!' I snarled." Disgust is the secret of lasting friendship for Thompson, who believes it's the major force behind human attraction and political power.

For all the craziness of American life in the 1960s, the nightmarish brutality of the twentieth century finds its best expression in real wars and their aftermath. John Hawkes's novel *The Cannibal* is set in World War II. It reaches back to the first World War and forward to predictions of a third one to create a sense of the cyclicity and inexorability of genocide and murder. This novel of horrendous atmosphere opens with an American in charge of most of devastated Germany riding through his domain on a motorcycle. He carries a sack filled with "unintelligible military scrawls" and has only "hypothetical lines of communication." Intellectual confusion, the devastations of Western culture, the sheer silence through which the American rides and to which he can bring no relief serve to make the country vulnerable to another Hitler.

The American as lonely hater, as impotent in victory, is one of Hawkes's targets. Zizendorf, the American's antagonist, is a German tyrant who will be the next Hitler and the creator of a third world war. Perceiving that the American is unable to be more than an extension of the devastated world, Zizendorf outdoes him easily and fills the country with his propaganda, his noise. Hawkes uses Zizendorf's political "certitude" to pave the way for the return of the entire country to an insane asylum. The asylum, ironically, does not consist of the schizoid confusions of the "real" world, but of Zizendorf's authoritarian platitudes. Insanity, in Hawkes's novel, is political faith.

The evil at the center of politics is not political but is a pathological thirst for control and domination. Hawkes has praised the novel which can "objectify the terrifying similarity between the unconscious desires of the solitary man and the disruptive needs of the visible world." He treats the tyrant as a gross magnification of everyone's love of disruption. His Zizendorf may come to power in a nation whose people are cannibals who literally dismember the young. Hawkes's tyrant is a devil whose rule will look like a black mass.

Hunter Thompson is about as eerie as a Peter Max poster. But he too accepts bad guys as our political reality and young, battered revolutionaries as their angry mirror image. In *On the Campaign Trail* Thompson says, "there are only two ways to make it in big time politics today: one is to come on like a mean dinosaur with a high-powered machine that scares the shit out of your entrenched opposition . . . and the other is to tap the massive, frustrated energies of a mainly young, disillusioned electorate." For Thompson the country runs on frustration, its power source is anger, its most marketable commodity is hate.

Hate freaks are among the newest American entrepreneurs. Alice Cooper, the bisexual rock star, grossed seventeen million dollars in one year by singing about sexual fury while holding a writhing, live snake between his legs. He climaxed a song about dead babies by finishing off a baby doll so realistically that two girls who watched the show tripping on acid went into shock. Asked what he intended to do about this, Bob Greene his former publicity manager, reports, Cooper answered, nothing. Nobody asked them to see such a sick show. Cooper had sensed that the audience for degradation was growing and could be won at the edge where life is only violence. His audience is young—largely people between fourteen and eighteen who have been brought up during a period of open rage between the sexes and justifiable outrage at social institutions.

Anger has brought the underground up and everybody down. Once the S-M spirit flourished only in the back-alley brilliance of William Burroughs' writing. Now editors of the most fashionable magazines print Burroughs' *advice* about social problems. Capote once shocked people. Now he plays the *enfant terrible* on talk shows. Selby achieved an extraordinary revelation of those anonymous haters, the as yet unknown Sirhans or Bremmers who are the excretions of American life, whose radical personal failures add up to a comment on how rabidly we are breeding the envy unto death. Sophisticated Barthelme taps the anger of the cosmopolite, the audience that demands elegant ideas and will accept them even in the form of an attack on the value of thought.

Anger is our leveler. Between underground fury and Olympian spite is the discontent of the audience. The reader quiet and desperate in his armchair has helped high culture, pop culture, and the underground to coalesce. As the audience feels ever more cruelly ground up by life, it becomes harder to shock and yet more eager for experiences that permit its anger to surface. The electrifying experience of cruelty can kill the audience's depression. S-M fiction brings the reader out of his down by making a spectacle of the mental see-saw between humiliation and rage. It offers the kick of the murder, the vicious political fight, the guerilla war, the spleeny vilification. What the reader gets is the literary equivalent of the lynching, the witch-burning that have both had their place and time in American life. What the audience finds in S-M books may be itself, out-grossed.

FICTIONS FOR SURVIVAL

If you can manage it, return with a portion of your weaned and grown-up feeling to any one of the things of your childhood with which you were much occupied. Consider whether there was anything at all that was closer, more intimate, and more necessary to you than such a Thing. Whether everything—apart from it—was not in a position to hurt or wrong you, to frighten you with a pain or confuse you with an uncertainty. If kindness was among your first experiences and confidence and not being alone—are you not indebted to it? Was it not with a Thing that you first shared your little heart, like a piece of bread that had to suffice for two? . . .

Rainer Maria Rilke,
second lecture on Rodin

John Barth is the brightest mark of a cultural faith in the self as a fabricated thing, a movable objet d'art. Barth's dazzling humor lights up an audience of role-players by taking the self-game one step further. Barth parodies how much people become actors, liars to avoid knowing who and what they really are. The fun comes from his exposure of the intimacy between quick-change artists and stable horrors. His comic faith is that if you see your defenses plain, you can build them higher; if you see how much you want to avoid what you feel, you can avoid it better. No Freudian, Barth is all for keeping life's surfaces lovely by keeping torment buried, by claiming that psychology is esthetics. Barth outdoes the audience in seeing happiness as the well-wrought barricade.

As Odysseus outwitted the Cyclops by claiming he was "no-man," so Barth's people outwit adversity by denying or erasing everything but a wily voice. And Barth's voices are always storytellers weaving fictions, tales, myths—all those glorious lies that can save them from themselves. Barth writes holistic fiction. For his people style is life-style, the logic of the sentence refutes the mind's chaos, syntax can be a saving grace and parody is the only force for order. Barth has made of stories and style a shield, of wit a sword, of diction a prime defense, of language a force that makes the self cohere and come alive. For Barth's characters are caught in the primal experience of nobodiness, the sense that they are blank as interstellar spaces, or selves so protean that they are nothing but pure possibility. But they find the magic words, the abracadabras, the shazams, the fables that ward off Mother Silence.

From the farces on current life—*The End of the Road, The Floating Opera* and *Lost in the Funhouse*—through *The Sot-Weed Factor* and *Giles Goat-Boy,* those historical epics, to the mythical heroes of *Menelaid* and of *Chimera,* Barth's no-men run back through time to legend, to supreme fictions. Barth turns the audience's obsession with "identity," its mania for the question "Who am I?" into a belletristic circus, making, on the great arena of history, a pure entertainment of falling apart.

Barth's heroes never ache or agonize, but disintegrate with the grace of virtuoso writers. For Barth is the grammarian of emotion. His characters move through the most terrible situations going dead to their pain as they become living, stylizing voices. As Jake Horner in *The End of the Road* remarks, "Articulation! there was *my* absolute, if I could be said to have one. . . . To turn experience into speech—that is, to classify, to categorize, to conceptualize, to grammarize, to syntactify it—is always a betrayal of experience, a falsification of it; but only so betrayed can it be dealt with at all, and only so in dealing with it did I ever feel a man, alive and kicking."

For Barth's moderns, life depends on words, codes, falsifications that will fabricate a heart. Horner is Barth's prototype of contemporary man, paralyzed in choicelessness because he feels nothing for anything. Having passed his master's orals at Johns Hopkins, Horner checks out of his room, for he has no reason to stay in it. He goes to the bus terminal to go anywhere twenty dollars will take him. But while making up his mind where to go he "simply ran out of motives as a car runs out of gas. There was no reason to go anywhere. There was no reason to do anything . . . except in a meaningless metabolistic sense Jacob Horner ceased to exist altogether, for I was without a character, without a personality: There was no ego; no I." He sits immobilized in the terminal for twenty-four hours. How to move a psychic paralytic? What to do when you do not exist? Barth offers a doctor whose therapies are all imperatives.

"Choosing is existence," insists the doctor.

> To the extent to which you don't choose, you don't exist. Now everything we do must be oriented toward choice and action. It doesn't matter whether this action is more or less reasonable than inaction, the point is that it is its opposite. . . . It doesn't matter whether you act constructively or consistently so long as you act. . . . Don't let yourself get stuck between alternatives. If the alternatives are side by side, choose the one on the left; if they're consecutive in time, choose the earlier. If neither of these applies, choose the alternative whose name begins with the earlier letter of the alphabet. These are the principles of Sinistrality, Antecedence, and Alphabetical Priority. There are others.

Unconcerned with emotions, disdainful of introspection, the doctor insists that people require rules, tight formulas, or roles to hold them together. What they need is his "Mythotherapy," which, he says, "is based on two assumptions: that human existence precedes human essence, if either of the two terms really signifies anything; and that a man is free not only to choose his own essence, but to change it at will." If Horner cannot exist in and for himself, he can be the parody of another. Play roles, assume masks, insists the doctor. "Don't think there's anything behind them, there isn't. Ego means I and I means ego and the ego by definition is a mask." The doctor's talent is for turning life into drama, for infusing the awareness of fiction into people who could never make it in life, but might thrive as art forms.

Barth's cheeriest spokesman for the glories of detachment, the doctor loves people as protean storytellers, performers who cannot feel a thing. But both Horner's disease and the doctor's cure are defensive maneuvers, designed to block feeling, to deny the continuity of personality and to focus on the forms of paralysis so that the content can disappear. What that content is is scarcely apparent to Horner, who goes dead to his own depression. But it is ominously suggested in Barth's descriptions of a plaster bust of Laocoön, that priest to whom Apollo granted knowledge of the contents of the Trojan horse. Driven to warn his people by his love for them, killed by a vindictive god whose passion is wide enough to annihilate him and even his innocent sons, Laocoön is the man cursed by knowledge and feeling alike. Horner's bust of Laocoön serves as a fixed point of consciousness, a death-mask signifying that intelligence, involvement and concern can end in the annihilation of the country you love, the children you want to protect, and the betrayal by the god you have served for a lifetime. Laocoön is the book's tragic presence, arresting Horner's attention, an omen that involvement unleashes destruction.

Barth turns *The End of the Road* into a tragi-farce on philosophy and infidelity, a parody of the agitations of love and thought. Horner, newborn mythotherapeutic poet, a pragmatist, neutral before all grand designs, is played off against Joe Morgan, the super-rationalist, idealist and advocate of nothing but absolutes. And between them is the honest, anguished Rennie, Joe's desperately loving wife, who has looked "deep within herself and found nothing," and who clings to all of Joe's ideas for the sake of wholeness and stability. The ménage à trois that ensues rises to a great spoof on self-consciousness of any kind, a parody of people who think they are saved by ideas, but who are doomed in Barth's vision of all relationships as deadly. Rennie, made pregnant by one of the two men, cannot endure the possibility of bearing Horner's child and resolves to kill herself. Horner, his disengagement breaking down, frantically gets his doctor to perform an abortion, and, in one of Barth's rare queasy scenes, he watches Rennie die on the operating table.

"Terminal," Horner tells a cabdriver. Will Horner go back to immobility? Neither in nor out of life, without even an "I," will he elude the destructiveness embedded in love or even in detached affairs? What stands at the end of the road is Barth's sense that peace and innocence may be found in paralysis, the well of choicelessness where the numbed mind and non-being free you from being alive. Why feel at all when all there is to feel is remorse? Why get involved when involvement is lethal? What seeps through every perfect line of the novel is Barth's sense that all anyone can do is mock his hang-up or control his collapse with witty detachment. Beyond the mockery is real life, the depression Barth deepens toward the ultimate choice, the terminal question, Should I live or die?

The Floating Opera is about Todd Andrews' decision and indecision on the day he decided to kill himself and changed his mind. No parodist, no mythopoet, Andrews is Barth's attempt at a character riveted in obsessive reasoning, in stating and negating, doing and undoing himself. A lawyer with a legalistic soul, Andrews spends ten years investigating the facts of the day he decided to die. And the fruit of this Inquiry is one logical proof whose final term is "There is . . . no 'reason' for living . . . (or for suicide)."

Formalized despair, logic at the edge of craziness permit Barth's characters to spring out as benign, witty presences, spinning out their existence on the thin line between paralysis and suicide. Andrews reaches the neutrality of a Horner. But Barth accomplishes a sublime spoof on the whole logical process, the power of mind to synctatify, order and categorize life into meaning anything at all. For beyond Andrews' survival theorem is Barth's vision of life as broken episode, the floating opera you can watch from a riverbank, never seeing more than the bit played before your particular spot. This sense of life as fragmented, elusive, incapable of arousing any gripping, coherent feeling, controls all Barth's realistic work and informs his conception of character as a compilation of ideas, roles, terms. For Barth the modern world is a game in pieces and human nature the prime jigsaw puzzle. What would the puzzle look like assembled? How does the human opera float? Lethally, according to Barth.

What human situations break through Andrews' logical screen are full of wrath. Andrews is third in a ménage à trois with loving friends who have adopted him. But after fifteen years of amiable sex with the beautiful Jane and friendship with her husband, after having possibly fathered "their" daughter, Andrews refuses to adopt *them*. Even as their lawyer he is so malicious that he does not know whether he will bother to win (as he can) the case over Mack's father's will, in which the Macks would inherit three million dollars.

Detached destruction is Andrews' forte. Only once in the Great War was his aloofness broken, when, overwhelmed by fear in a trench, waves of terror washing over him and loosening his sphincters, a German soldier sprang into the trench, equally terrified and loose-boweled. Clasping each other through the night, they became fiercely protective of each other, each looking out while the other slept. But towards morning Andrews became suspicious—what if all this neediness was only a pretense? What if the German was only waiting to kill him? He drew his bayonet and killed the man who, roused by the knife from sleep, looked at him with shocked, astonished eyes and died. The urge to kill any closeness in any encounter suffuses Andrews' life. For his alternative to detached cruelty is involved cruelty. Just as he destroys the trusting soldier, so he needs to demolish the Macks' affection for him by outraging them with phony tales of his sexual exploitation of the poor black girls of the town.

How to explain the urge to wreck any involvement? In *The Floating Opera* Barth begins to chronicle its history. Left motherless at seven, Andrews is raised by a succession of maids and is nodded to occasionally by his neat, close-mouthed father. At seventeen, infatuated with a worldly local girl, he eagerly climbs into bed with her, but, catching sight of their coupling in the mirror, he bursts into laughter at the ridiculousness of sex. Later, having joined his father's law firm, he comes home to find his father impeccably dressed but hanging from a cellar beam, his neck snapped. Andrews, unlike Horner, has a past, a family, a history of emotional loss, thwarted love, outraged sonhood, sexual detachment. An emotional paralytic, he is doomed to emptiness, forced to devote his life to an Inquiry that lists only what he did and not what he felt on the day he decided not to die. Like Horner, that comic existentialist, he is nothing but the record of his decisions and indecisions, nothing but the sum of his choices or choicelessness. Andrews is finally saved not by logic but by paralysis, by his inability to feel enough joy or despair to want to live or die. The final term of his proof is not only Barth's mockery of logic as the syntax that can make human nonsense cohere, but a bitter recognition that logic, like mythopoetry, does not even apply.

> What springs from Barth's novels of contemporaries is ridicule of both the disease of paralysis and its "cures." Horner and Andrews are technicians of collapse, craftsmen organizing their decline without understanding it or changing it at all. For neither the detached imagination nor detached reasoning can put the human puzzle together. Could introspection do it? Barth's one attempt to explore the innards of the mind is *Lost in the Funhouse,* a collection that adds up to a portrait of the writer as psyche. Barth virtually attacks introspection as the final dead end, and personality as nothing but insecurity or nagging complaint. Ambrose M., the young artist in the title story, cannot prevent his narrative or his life from going nowhere. Divorced from that rising action, climax and denouement that are Barth's happiest metaphors for sexuality, life and art, Ambrose M. cannot reach the fun parts and so endures all the frustrations of sex and language, blocked from the lover's and the writer's climax. This story is Barth's fugue on the nether regions of the modern self and modern literature, on the torture of character and language in the back closet of the mind where the artist, trapped within his own sensations, is no more than the echo of his own ungrammaticized, unformalized neurosis. As Ambrose grimly realizes from the depths of the mirror-maze, "Even if you had a periscope, your eye would cover up the things you really wanted to see." For the writer who sees nothing but his own eye has the most boring, most painful modern disease. Barth equates introspection with private fear, with inadequacy in all its humiliations reflected through the mirror-maze of consciousness. And rolling through his novels of contemporaries is a relentless indictment of philosophy, logic and the ruminative process as dead ends in the funhouse. How to get out of the terminal maze of modern times?

Apollonian, impersonal, cool, Barth is impelled to get out of the paralyzing trap of personality, to find the Pegasus that can leap into Art. And History is his flying horse, the inspiration that effects the greatest distance from present despair, that offers the dream of possibility, of youth, of a childhood full of live nerves, live wonder, a time when fictions were as real as facts, and the funhouse was fun. In *The Sot-Weed Factor*, one of the extraordinary literary feats of the sixties, Barth wrote of seventeenth-century America as a no-man's land whose sheer wildness generates the "freedom" that "makes every man an orphan." For Barth's characters reemerge in the past as Heroes, as surging historical phenomena. Personifying the American continent is Henry Burlingame, abandoned at birth in an open boat, half "salvage Indian," half English Lord, one-time gypsy, scholar, and one of those violently polysexual men whose tastes extend through both sexes to shoats. For Burlingame is not paralyzed by choice; he chooses everything at once. Disguising himself as Lord Baltimore one day and a pirate revolutionary the next, he works on both sides of the cause. Be yin, be yang, be Lord and "Salvage," implies Barth, be anything but yourself. For nothing excites like mutability, nothing like losing one's own emotion, one's own past in the national energy, the national history. Burlingame multiplies himself out of the obsessional, paralytic bind of Horner and Andrews, out of human stature into a living demonstration of Flux.

To be human is to be paralyzed; to have a purely personal history is to be crushed by the weight of its disappointments; to be all Lord and no Salvage is to have no sex—these are Barth's obsessions. To underline them he sets against Burlingame the young Anna and Eben Cooke, twins left motherless at birth and made virtually fatherless by the quirkiness of their provincial English father who makes Burlingame their tutor. Burlingame lusts after both, excited by twinship as primal sexuality, the coalescence of yin and yang. But for Eben all this intimacy only breeds abstinence and total immobility. Self-styled Virgin and Poet, Eben at Cambridge "did not even deign to dress himself or eat, but sat immobile in the window seat in his nightshirt and stared at the activity in the street below, unable to choose a motion at all, even when some hours later, his untutored bladder suggested one."

Burlingame goes on, plugged into the immense energy of nascent cities, "salvage" bands, of a continent infusing its wildness into everything. For the American past is Barth's dream of Odysseus as Cyclops, one fusion of mind and power. And with his absolute energy, fragmented sexuality, and diffused personality, Burlingame is the first Barthian hero to invent "no-manness" as a national ideal, to turn the chaotic, the disintegrated heart and the broken self into the best defense against paralyzing depression.

Infusing all Barth's later work is a passion for the single anomalous figure who depersonalizes, abstracts and enlarges emotion toward a heroic dimension. In *Giles Goat-Boy* Barth attempted a Messiah who might be the perfect American savior: a hybrid of a machine and a goat. Fathered by a computer, born of a virgin who immediately abandons him in a library dumb-waiter, made lame in the chute, raised as a goat among goats by a scientist soured on humankind, the Goat-Boy, Barth insists, follows the archetypal pattern of heroes, and will certainly grow into the Grand Tutor who will decipher what is "passed" and what is "flunked" in a universe run like a university. Overlong, belaboring the analogy between studentdom and humankind, *Giles Goat-Boy* is probably Barth's least successful novel. But it powerfully conveys his new way of dealing with that recurring figure in his work: the damaged, abandoned child, the inhuman man who cannot feel a thing.

Even this weak novel shows Barth's genius for archetype as analogy, as the means of expressing, ennobling and enlivening the anguish of the man paralyzed by depression. As boy the Goat-Boy is another Eben Cooke, another Andrews, another Ambrose M., another of those emotionally worn-out children. As goat he has lustiness without anxiety, no introspective bent, no awareness of depression. As Goat-Boy he fuses his experience above the human line and becomes a Comic Hero. For the answer to depression is entertainment, the fictions that for Barth do more than lighten the heart. Through analogy after analogy, Barth builds an emotional system where parody is the bridge over the feelings you have to the ones you do not, where it is possible to estheticize yourself into one of those beautiful people who are history's adornments.

Barth's heroes are not necessarily happier than his moderns. But they all feel more "important," more able to dissolve their insecurities in an idealized image, to alchemize their own emotions. Barth's hilarious *Menelaid* is about a man caught between reality and art, the fact and the poetry of himself. Can he do the alchemical trick? In youth, Barth's Menelaus is all too aware he is a nobody, "less gooded than Philoctetes, less brave than Ajax, less crafty than Odysseus . . . . Why then did Helen choose him of all the great men of Greece? To her Menelaus signified something real, as Helen him. Whatever was it?" Such questions presumably drive Helen to Paris. Eleven years later, Troy in ashes, Menelaus grips Proteus to get an answer to his woes. But entwining Proteus through all his transformations, Menelaus loses the spirit of inquiry. "Helen loves you without cause; accept her without question," the Old Man advises. And Menelaus does, even while Helen insists: "Husband, I have never been in Troy. . . . What's more I've never made love with any man but you. . . . Before Paris could upend me, Hermes whisked me on father's orders to Egyptian Proteus and made a Helen out of clouds to take my place."

Cries Proteanized Menelaus:

> Presently my battle voice made clear my grown conviction that the entire holocaust at Troy, with its prior and subsequent fiascos, was but a dream of Zeus's conjure visisted upon me to lead me to Pharos and the recollection of my wife—or her nimbus like. For for all I knew I roared, what I now gripped was but a further fiction, maybe Proteus himself, turned for sea-cow respite to cuckold generals.

Is Menelaus Proteus? Helen Proteus? Proteus Proteus? No matter, in this domestic comedy, or in all Barth's mythic tales where the only answer to infinite despair is infinite credulity.

Marriages, like art, survive through that suspension of disbelief, that storytelling that in Barth's work always involves a suspension of real emotion, an almost willful abandonment of pride and resentment, tenderness and need. What Barth's Apollonian art demands is precisely the collapse of the personal, emotional, introspective. With myth, with legend, with folklore, Barth has worked toward covering the face of human experience with classical dramatic masks so that characters in his stories achieve in their own lives the distance between storyteller and story, player and play. Survival in Barth's world requires that emotion be stylized, that feeling be converted to yarning. For to feel is to lose—lose Helen, lose one's way in the funhouse of life and art, lose oneself in the labyrinth of personality. And no one has written with more distaste of the private, the personal demon; no one has found a more elegant alternative than Barth in that sublime name *Chimera*, the legendary beast only Pegasus could corner, and that only Barth could make the emblem of every writer's private terror.

What to do if your Pegasus has grown fat? If inspiration no longer soars, no fables come to mind? The three novellas of *Chimera* are all about creeping silence, about heroes in middle age who are terminal mirror-maze sufferers. By far the most brilliant is the *Dunyazadiad*, in which a Genie who seems to be Barth appears to Scheherazade and her sister Dunyazade and announces himself as a writer gone quiet: "My name's just a jumble of letters, so's the whole body of literature: strings of letters and empty spaces, like a code that I've lost the key to." He wishes "to go back to the original springs of narrative" for inspiration; "and of all the storytellers in the world his very favorite was Scheherazade."

But Barth's "Sherry," though she wants to stop the Shah from deflowering and murdering virgins as he has done for years, knows only stories that already exist and has no idea of how they could be told to the Shah. Each day the Genie appears to tell her, from the future, stories she is said to have written in the past. Each night, after making love with the Shah, she tells them to him, leaving the climax for the morning in the hope that he will let her live another day, her little sister Dunyazade watching her lovemaking and storytelling from the foot of the bed. At the end of the Nights the Shah marries Scheherazade and marries her sister to his brother, all amid great rejoicing. But Barth's tale is a comment on how feeling would make this resolution impossible. He confronts the tragic split between life and art, the terrible way imagination betrays itself to emotion. For who could rejoice who had to perform to stay alive? What gratitude has an artist who has just gone through a thousand and one one-night stands?

For Barth's people, in the silences of love and storytelling, in the relation between lovers, artists and audiences, is always the fact of destructiveness, the impulse to kill each other. Sherry plans that she and her sister will each cut off her husband's "bloody engine" and "choke" him on it. How to save Scheherazade from being real? How to save the bridegrooms? The *Nights?* In this tale of story within story within story, how else but with still another story? The Shah's brother beguiles Sherry's sister with a tale of love and a plea to love "despite her heart's feeling," to make love "as if" she loved. Never has Barth so plainly put the imperative to live against the grain of basic hate, never so pleasingly argued for parody as the only force for grace.

Barth is the parodist sublime, the writer of parables against the betrayal of beauty by feeling, of myth and even mythotherapy by reality, of imagination by human need. Identityless as the Scheherazade of the Nights, his characters are supreme articulates, are nothing less than voices for fiction. Grappling toward their own survival, against the tide of their own inability to will, or to care, Barth's people nevertheless celebrate the images of heroes and lovers, preserved in legend through centuries of real frustration.

From *The End of the Road* to *Chimera* Barth has spoofed the search for identity into an intense, serious pursuit of a mythology of emotion. He adores parody as self-definition and self-defense, the means of saving oneself through reverse mimesis. Sherry and Donny abandon their rage by faking love, encapsulate themselves in fiction and bow neatly back into the *Nights.* Nothing is not parodic in Barth's world, neither storytellers, nor enchantresses, nor stories that cohere only through their relation to other stories, an existing myth, or an archetypal pattern. Happier than reality is Barthian similitude, the power to live "as if" you were another, that transformation that turns haters to lovers and is foretold in the magic formula of Scheherazade and the Genie: The Key to the Treasure is the Key. Magic words are the Key and the Treasure; storytelling is life's means and only prize. And no one has written more glitteringly than John Barth of the worthlessness of the heart, or the great munificence of language in bestowing so much grandeur, so much richness, so many pearly epigrams on all us swine.

V

THE VICTIM IS A HERO

Only the sacrifice of an innocent god could justify the endless and universal torture of innocence. If everything, without exception, in heaven and earth is doomed to pain and suffering, then a strange form of happiness is possible.

Albert Camus, *The Rebel*

The captive male has traditionally been held by family responsibilities, by social class, by work. But in many recent novels the hero finds his cell is made of women—his mother, his mistress, his wife. His chains are his often mysterious infatuations. For such heroes life means being pinned between anarchic and holistic impulses continually expressed in sexual ambivalence, in a perverse drama of sexual power and victimization. In these tales of male susceptibility, the hero is a prisoner of sex who practices the politics of vulnerability.

No one has done more to explode male freedom as a myth than John Updike in novels of American men whose lives, from cradle to grave, are structured by women. His characters are often the philanderers, the printers, the ministers, the businessmen who seem freewheeling until the press of Updike's intelligence reveals them to be captives in that velvet glove, the female presence. These Adams were beckoned by Eves to fall from the grace of maleness into a life of labor. God reveals destiny to man, you might say, through marriage.

Updike's destiny as a writer is to see the world, for all its tensions, as a wedding; to see a man's wife as his fate. He is one of the few writers to take from modern discord the gift of reconciliation, to find even in unhappy marriages necessary harmonies. Updike's novels use marriage as the index of social cohesiveness, of economic, political and sexual change. They register the impact of the revolution that is changing how people are with each other.

Rabbit, Run (1960) and *Rabbit Redux* (1971) show transforming American Moms and Pops. In *Rabbit, Run,* Momism is flourishing in an American housewife who is controlling, ironic and relentlessly involved with her son, whose narcissism she alternately wounds and strokes. Rabbit's Pop is a stereotyped, hardworking man who is passive before his wife. This pair does not reproduce their kind. Rabbit has had a taste of abundance as a high school basketball star. At twenty-six, he is married, aged by his heavily pregnant wife who drinks too much, by his concern for his two-year-old son, by his humiliating job and by his fear of becoming like his father. He runs away. His sister, Mim, won't be a kitchen martyr. She becomes an efficient manager of people, a call-girl who thinks life works best on a hard-cash basis for self-protective independents. These are Updike's new Americans: the man engulfed by self-destructiveness and anger, the woman who is all cool sex. What produced the mutation?

Updike is the first fine writer since D. H. Lawrence to define maleness as sexual and economic responsibility for women, and to connect the decline of society with the decline of masculinity. *Rabbit Redux* offers an ambitious patterning of the social forces that he feels create uncertainty—automation that devalues work, Vietnam protests that devalue authoritative institutions, sexually free wives who help elevate the values of self-expression over self-control and make the working man's pro-war, pro-family, pro-patria values seem outmoded. All these make Updike's men feel obsolete.

At thirty-six, a hardworking linotyper, Rabbit would seem to have become his father. But it is no longer possible to live his father's life. He is replaced at work by a photo-offset machine and in his wife's bed by a man who is an infinitely better lover. Technology has changed his mother's life, too. Old and ill, she is propped up and pepped up by a pill that brings her sexual fantasies she presumably never had as a healthy woman. Is sex the medicine of the sixties? Only for women.

For Rabbit there is no tonic for frustration but revenge. He takes in a black nihilist and a girl who has dropped out of upper-middle-class life into drug addiction. He lets the man abuse him and takes out his rage passively by letting him kill the girl. His destructiveness and self-destructiveness get his house burned down. Updike brilliantly compresses into the fall of a single house and marriage the war between the sexes, the dissidence between classes and races, the efficiency economy that undermines a man's sense of himself. Houseless, jobless, wifeless, Rabbit crumbles. Giving up on being a father to his son, he goes back to his Mom.

What Updike bares in his beleaguered male chauvinists are the forces that keep men riveted on women as the solitary source of meaning in life. Rabbit and Janice are a matched pair whose collusive entanglement with each other, however unhappy, is right. When Rabbit runs out on his wife, Janice drunkenly drowns their infant daughter. When Janice runs out on Rabbit years later, he lets his young girlfriend die. Neither Janice nor Rabbit is happy with the other, yet the presence of her husband helps Janice function and the presence of his wife helps Rabbit behave like a man. Janice is the control on his anger, his self-hatred. Why do men need women to structure their lives?

Updike puts life together as a sophisticated Oedipal knot in which a man is tied at both ends. His characters fear being in control, in charge, but are equally afraid of being suffocated and controlled. Their inhibitions bring them the worst as sons and lovers. Lawrence's famous son thought he would be a lover when his mother died. Updike's characters keep their mothers alive forever by remaining in the box of contempt and possessiveness that is the mother's personality. They fall for women, but cannot stop hating them or turning bliss to ashes. When Rabbit runs out on Janice, he meets the perfect companion: Ruth is a good-natured woman who lives by her body. Her sexual experience makes Rabbit feel as excited and competitive as he did on the basketball court. He wants to oust all other lovers from her mind. He succeeds. But he cannot enjoy his high for long.

Rabbit dreams he is back in his mother's kitchen with his sister Mim. Mim opens the icebox door and they see a block of ice in which something that looks like a heart is frozen. His mother begins scolding Mim for opening the door. Mim turns into Janice; her face melts. But it's Rabbit who is furious at women for trying to melt him. He needs to keep cool, not only to preserve himself in his mother's kitchen, but to ice his destructiveness. That Janice melts into oblivion reflects his sense that to lose your cool is to die. Sister, wife, mistress are interchangeable as women who are in danger if his feelings are freed. What might work for him is a woman who remains detached, aloof, who promises a warmth she won't deliver. This is the dark side of sex where idealizing, romanticizing a woman is better than having her because intimacy unleashes destructiveness. Ironically what makes Rabbit feel alive is the total conquest that makes him want to destroy.

At their worst, Updike's men are victims of forces which he understands but they do not. They are unwilling to thaw or to change and unable to remain the same. They think they want a warm woman, but can only live with a cold one who keeps their contempt alive and their anger in check. They hate their mothers' power to paralyze them, but cannot get over the sense that only their mothers truly love and understand them. They look to mistresses and wives to be stronger forces than their mothers, powerful enough to free them. They identify with the children they father, wishing they could be born anew and mothered by their wives. But despite their hopes, getting into a woman means getting back to their mothers' kitchens where there is guilt and frustration for them and hate for any woman who threatens to melt the iced anger that binds them to their mothers.

How this bond operates is one theme in Updike's fine short novel *Of the Farm,* in which a man, against his will, finds himself looking at his life through his mother's disparaging eyes. When Joey Hofstetter and his new wife arrive at his mother's farm, he watches his sensual wife, whose body he loves, walk toward his mother. Knowing his mother will see her only as a large, gross, painted woman, he develops a kind of double vision, seeing her that way, too. He begins to belittle her, he complains about her "promiscuity" with men before she knew him; he hates himself for being fool enough to marry her. As his mother, who belittled his first wife during his marriage to her, insists, wife 2 seems not so bright, charming or attractive as wife 1. Joey's self-hatred turns to hatred for the wife who is the living sign of his bad taste. His fate is to need the contempt and self-contempt that keep him bound to his mother.

Can fathers show sons the way out of ambivalent love? What has changed are not the problems of fathers and sons, but the solutions. In *The Centaur,* the father, when he tells his wife of the terminal sickness in his gut, finds that she only becomes enraged that he may cop out on her by dying. She goes on to criticize him for being cheap and afraid of sex. This father has dealt with great provocation and great impulses toward flight by paralyzing himself and involving himself in his son. His sacrifice of his life is so large and so painful that Updike buries it in the mythological story of the centaur. The novel deals with the son's tender, compassionate perception of his father's vulnerability not simply to death, but to the unfair abuse of the high school students his father teaches, the principal who tyrannizes over him, the passers-by he befriends. His son sees him as "gathering things to himself and letting them drop," as unable to care for or protect himself adequately, as a victim. Through the abstraction of the centaur's story, Updike suggests the father is a universal predestined male victim; his beginning is the end of him.

Chiron's story may be emblematic of what happens between mothers and sons, husbands and wives. Surprised by his snooping wife Rhea, Kronos, in the midst of making love to Philyra, changes himself into a stallion, leaving his seed "to work its garbled growth in the belly of the woman." She bears a centaur, half man, half horse, and abandons him in disgust, pleading with the gods to release her from her shame. She is turned into a linden tree. Her infant starves. As a man he takes a manly, compassionate view of his mother's pain, he sees himself through her eyes. Yet

> when he contemplated the fable of his birth an infantile resentment swelled up bitterly within his mature reconstruction, the undeserved thirst of his first days poisoned by his mother; and the tiny island on which he had lain exposed seemed the image of all womankind: shallow, narrow and selfish. . . . Too easily seduced, too easily repulsed, their wills wept self-indulgently in the web of their nerves, and they left their dropped fruit to rot on the shore because of a few horsehairs. So, seen through one side of the prism he had made of the story, woman was to be pitied; and through the other, to be detested. In either case, Venus was reduced.

What Updike embeds in this story is how primal and engulfing male hatred of women can be. Chiron/Caldwell hates them for their vulnerability to sex, their invulnerability to him, and for escaping the net of responsibility in which he is caught. His frustration seethes beneath a thin veneer of protectiveness.

Chiron, mortally wounded when accidentally hit with a poisoned arrow meant for someone else, is the slaughtered innocent as adult male, wounded but doing his best to heal others until the pain of his efforts "kills" him. Mr. Caldwell rises above his life, is more than his life by virtue of his extraordinary self-sacrifice, his sense of the value of self-renunciation for his son. Updike makes his sacrifice older than Christ's, casting it in the figure of the centaur who passes through death to become the constellation Sagittarius, the archer who, having been shot, remains for all eternity poised to shoot. Is centaur-love the poised attentiveness to an interesting target? The search for a weak spot? What Caldwell has given his son is the centaur method of buried resentment and overt protectiveness.

Updike's heroes try to control their anger and not shoot the arrow. Many aim for Caldwell's solution, and do their best to find fulfillment as fathers. The father of four, Updike has written some of the tenderest, most realistic accounts of small children in recent fiction. He has no trouble making his small characters talk, walk like two-year-old troopers, get carried sleeping to the bathroom, curl up with their bottles, weep over dead pets, sleep touching a sister's hair. Updike's men express their own vulnerability through compassion and protectiveness for their little children. But kids grow up, disappoint, in later novels become allies of an angry wife or competitors who do not rejuvenate their fathers but make them all the more conscious of age.

Rabbit wants to be reborn as a winner of unambivalent love. He looks for advice from his high school coach and finds a pathetic youth-chaser, a hideous parody of himself, a sixtyish man who has run out on his wife and fills up his life with basketball and prostitutes. Rabbit recoils from him and looks to his minister, the young Reverend Eccles, who tries to persuade him to go back to Janice. Eccles himself has an empty marriage in which his pretty wife is no source of pleasure. His gaiety, she notes bitterly, is spent elsewhere. She derides him for fearing sex, for avoiding her and working all the time. Eccles is faithful to her but only through a conscious, willed effort. A middle-class version of Rabbit, he makes the usual middle-class compromise, burying his needs in work and controlling his own impulses by controlling the impulses of his parishioners. He only strengthens Rabbit's hatred of responsibility. Neither fathers nor sons nor coaches nor ministers can show the way to the grace of maleness, or provide a sense of nascency.

Women are the only masculine pursuit in Updike's novels that offer the promise of rebirth. Rabbit's grace comes from his wish to be made new; his decline from losing his power to believe he can be. Rabbit is the dark side of every man who realizes that the best part of his life is behind him, that his idealism, his pride and power are gone. The responsibilities that were once the thorn in his side painfully seem the only justification for his life. The great irony in both Rabbit novels is the existential contradiction between what a man wants and what his heart will allow.

Rabbit is man as the scared animal who runs and reproduces, controlled by compulsion, by an economic structure that keeps him down, by impulses that turn him to lead. He feels obsolete as a person because he tried to take his father's path in a world that has changed all the road signs. Can the balance between happiness and tragedy be tipped by where you put your eyes? The beauty of the Rabbit books comes from Updike's compassionate perception of the strangling boundaries of working-class life. The brilliant opening of *Rabbit Redux* describes a world without options:

> Men emerge pale from the little printing plant at four sharp, ghosts for an instant, blinking, until the outdoor light overcomes the look of constant indoor light clinging to them. In winter, Pine Street at this hour is dark, darkness presses down early from the mountain that hangs above the stagnant city of Brewer; but now in summer the granite curbs starred with mica and the row houses differentiated by speckled bastard sidings and the hopeful small porches with their jigsaw brackets and gray milk-bottle boxes and the sooty ginkgo trees and the baking curbside cars wince beneath a brilliance like a frozen explosion. The city, attempting to revive its dying downtown, has torn away blocks of buildings to create parking lots, so that a desolate openness, weedy and rubbled, spills through the once-packed streets, exposing church facades never seen from a distance and generating new perspectives of rear entryways and half-alleys and intensifying the cruel breadth of the light. The sky is cloudless yet colorless, hovering, blanched humidity, in the way of these Pennsylvania summers, good for nothing but to make green things grow. Men don't even tan; filmed by sweat, they turn yellow.

Working-class "ghosts" in a "wincing" city, under a sky like a "frozen explosion," filmed by sweat and "clinging indoor light" pass from adolescence to old age as quickly as beer goes down in summer. Do middle-class men realize the dream of eternal youth? Not exactly, but in Updike's novels of middle-class life, the dream surfaces in the youth-cult values of the "post-pill paradise," the tennis matches, parties, days at the beach that are the blessings of suburban affluence. Updike's middle-class novels offer characters who are better able to control their situations, sexual heroes who have discovered laissez-faire as a life principle. Updike casts a light eye on the new world where everybody thinks anything goes. The Oedipal situation turns around as experience replaces responsibility as a value and Mom and Pop are available. Irony isn't psychological bile, but frothy social confusion.

Updike is hilarious on the ill manners of middle-class marriage in a town where the basic social unit is the man and wife who are sexually disenchanted with each other. In *Couples*, Updike's smart, popular novel, when Angela, oft-condemned by her husband for her repressions, begins confessing her interest in sex and asking for help, her husband becomes enraged. He is happy to be unhappy with her because he has learned how to make his contempt work for him. He regards her as that perfect ally, the wife who is a defense against life: she holds his hand while the irate husband of his mistress tells him off. The long-suffering Mrs. Marshfield in *A Month of Sundays* knows how to humiliate her husband's mistress almost as well as he does. Updike's wives and husbands are "the matched jaws of a heartbreaker." The payoff is for the husband who learns how useful it is to have a wife who loves him just like a mother.

Is life better when a wife is a sister in the hunt? Having escaped one dilemma Updike's men and women find themselves faced with another: sex, Updike implies, frees men for other things, but binds women deeply to the man who pleases them. After the most ecstatic experience the couple is often left upset: the man wants to please (Updike's men are gentlemen) but he knows he cannot give the woman what he thinks she wants (quaintly, it's marriage). The woman, whether someone's wife or not, who watches her lover rise, dress and run, feels left behind. She brings her frustration home.

Updike's men are helpless before their need to love and wound, powerful in their ability to sexually enchant. The father in *The Centaur* expresses his hostility to his wife by renting unheated houses with collapsing kitchens. Updike's sexual heroes use sex to tie women in knots. Like Reverend Marshfield they maneuver women into humiliating each other. Or they get caught between mistresses and wives who each want something, so that, as Marshfield puts it, "in their midst I was powerful and felt helpless." Their helplessness drips with their coercive, angry attempt to maneuver the women they hurt into offering them maternal protection. Cornered, they rhapsodize on the disparities between what the sexes want, declaring that woman-is-earthbound and man-is-stargazer. Their stargazing invariably involves a wish to look past female pain and their own sexual anger.

What fascinates Updike is how richly ill-suited relationships can still be a source of the sublime. This writer's defense against despair is style, the management of language to surmount debacles. What counts is not the disaster of human relations but the point before the fall, the patterning of life toward what he calls in an essay, "The Future of the Novel," "that eventual copulation that seems to be every reader's insatiable and exclusive desire. . . . Not to be in love, the capital N Novel whispers to capital W Western man, is to be dead." As Marshfield, a newborn writer, says to his beloved and pursued reader, the adulterous Ms. Prynne: "There was a moment when I entered you and was big, and you were already wet, when you could not have seen yourself, when your eyes were all for another, looking up into mine, with an expression without a name, of entry and alarm, and of salutation. I pray my own face, a stranger to me, saluted in turn."

Can love as salutation last? Updike once spoke of a dream he had at fourteen of a knight in armor who falls in love with a Polynesian girl and pursues her across the world. On her island he finds her, looks at her through a palm frond and dies. "I never wrote that novel, the historical one," he said. But Updike has written it as novels in which to look too closely is to diminish the ideal woman with her own reality, and to die a little oneself. Updike peoples his novels with men who regenerate themselves by not seeing the women they make love to. After the salutation they foil, belittle, betray, disappoint but never cease to involve and ignore the women who so imperfectly stand in for the force to which they have given their lives—the female principle, the lady of rebirth.

Updike is not a sexual materialist. He has a charmed vision of sex as the great reconciler, the beginning and end, the origin and the goal, the comforter of characters who fear that the dead part of themselves may defeat the living. Reverend Marshfield jokes that there is no more beautiful phrase than "sexual object." Updike's characters are so in love with their *objective* that they cannot help, within the limits of the bed, loving the women with whom they achieve it. How should women feel? Like Yeats's Crazy Jane? She confides:

> Though like a road
> That men pass over
> My body makes no moan
> But sings on
> *All things remain in God.*

Updike's happiest characters are those who do not lose the power to believe there are Janes who make it all possible.

Updike has an intense preserved idealism. He reads the world as a prophecy of something better. There is little spoken love in his novels, but love is there in the unsparing skill with which he creates people who continually absolve each other. If love is the power to live as though fantasy and fact were indistinguishable, Updike is blessed. He believes fantasies virtually create facts. "In my own case," he said in "Why Write?", "I have noticed so often it has ceased to surprise me, a prophetic quality of my fiction, even to the subsequent appearance in my life of originally fictional characters."

Updike is a great dreamer whose novels are nevertheless starkly realistic about those relations between the sexes that head toward antagonistic love. *Marry Me* is about a man who, bewitched between his mistress and his wife, finds his destiny is sexual calamity. This novel is a subtle exposure of what you might call tender malice.

Jerry adores his beautiful blonde mistress, Sally, who drops her children elsewhere to come to him on an idyllic beach. Breathing her pleasure in her lover's ear, she hears him softly inquire: "Do you mind . . . the pain we're going to cause?" About to make love to Ruth, his as yet unknowing wife, Jerry purrs, "Tell me about Sally." Each woman furiously believes the other is his favorite subject. Jerry uses even his death anxiety as a jealousy hook. To Sally: "I look at your face, and imagine myself lying in bed dying, and ask myself, 'Is this the face I want at my death bed?' I don't know. I honestly don't *know*, Sally." To Ruth: "Whenever I'm with Sally I know I'm never going to die." "And with me?" "You? You're death. I'm married to my death." But packing his medicated inhaler to leave Ruth, he interrupts his complaints with the sweet inquiry: "Shall I wait until you fall asleep?" Updike's humor captures the disparity between the dreamlike pleasures expected between the "lover" and his "mistress" and the hesitancy of the suburban swinger. His wit blooms in the gap between the euphoric outburst "Marry me" and the failure of those magic words to change one's soul.

The greatest love in *Marry Me* is Updike's. It is evident in his generosity to Ruth. Her acid accuracy about her husband's flaws is unaccompanied by any desire for revenge, any critical bent, any interest in throwing him out. Updike's women are mirrors of his appreciativeness of the patient forgiveness and excitement that his Janices and Ruths divide between them. And while Updike is grateful for the nobility of women, while he grants them every palm of virtue, he never seems to try to know how they might judge themselves for staying with a man who offers his arm, his hand in marriage, to help them fall.

The sign of Updike's grace as a writer is the superb clarity with which he renders the man bewitched by his ambivalences. Updike writes with hard beauty of men endlessly wandering the labyrinth of their own needs, irresistibly child-like in their faith in the magic of a woman and abusively fearful of broken spells. He writes without sentimentality but not without compassion. He brings to mind the contemplative exactitude about complex and paralyzing appetites that St. Augustine achieved when he wrote of himself: "Of a forward will was a lust made, and lust served became custom and custom sated became necessity. By this chain was I enthralled."

Is the clear wrath of the outraged husband better than the pain of a philanderer? Saul Bellow has written marvelously of men who are victims of women. Easily one of our best novelists, Bellow has a superb unrhetorical style, remarkable for its immediacy and color. His novels often display a speculative intelligence caught in sexual emotions like quicksand. This is the plight of the man whose great family feelings are the death of him, whose allegiances perpetually bring him down. The recurring problem in Bellow's novels is involvement. The appeal of his best recent characters is the richness of thought and interest they bring to a sense of sexual involvements as dangerous. Their intelligence offers the continual promise of a solution. Why is this promise never fulfilled?

Herzog, Bellow's stunning performance as a novelist, catches Herzog falling apart from rage over his wife's betrayal of him with his best friend. In his pain he feels impelled "to write letters, to explain, to have it out, to express, to set straight, to intervene, to put into perspective, to balance, to remedy, to justify, to confess, to atone." The verb list, Bellow's standard device for expressing desire, is so inclusive it obscures how much Herzog uses his mind for very different purposes.

Herzog is smart enough not to be smart about himself. He uses his intelligence as a weapon against the turmoil of his personal life. Ostensibly in pursuit of a great synthesis of Western thought, Herzog is best at manipulating and exploiting abstractions to break the connections between his ideals and his life in order to retain his self-love. Even the form of his thought, the letter, is a discrete communication which invites no response and gets none. The dense intellectual content of the novel and the immense charm of Herzog's sharpness about social philosophy serve as a smokescreen for a personality so essentially rigid and narcissistic it can encompass no negative or critical idea of itself. Enshrining an adorable self-image in seductive abstractions, Herzog permits his life to remain rigidly set on a course of destructiveness and frustration. This novel in which Herzog claims to look for universal truth is a documentation of the benefits a man derives from ignorance of himself.

Herzog is, of course, objectively the "wronged" party in the triangle, displaced by his best friend not only in the love of his wife, but in the affections of his child. But he uses his victimization to conceal his malice and get what he wants. The family values Herzog professes to feel again and again are values he attaches to his mother and brothers, not to any of the families he forms himself as a man. It is the family in which he was nurtured that he wants; it is to be a son, not a father. Not seeing or knowing himself permits him to think of himself as a man but still derive the benefits of being a child.

Selective vision is a technique Herzog learned from his mother. "Mother Herzog had a way of meeting the present with a partly averted face." He applies her method to getting what he wants by appearing to be stupid. One of his childhood memories is of a cold day in January when his mother pulled him on a sled and was met by an "old baba in a shawl" who said, "Why are you pulling him daughter! daughter don't sacrifice your strength to children." "I wouldn't get off the sled. I pretended not to understand. One of life's hardest jobs, to make a quick understanding slow. I think I succeeded," thinks Herzog.

Not knowing himself gets Herzog off the hook of responsibility. The failure of his marriage permits him to win his otherwise unobtainable goal—to be cared for, protected, and worried over by his older brothers, who turn up to lend money, bail him out, give him advice. Even his sense of failure as a man revives the warmth of his Depression childhood. His disappointment with his wife stirs up memories of his mother and his love for her beautiful, suffering face. His first wife, whom he describes as a moody but "dependable Jewish woman," was like her; for his second, he marries the aggressive side of himself and finds he cannot live with it. (He complains that his wife edged him out of university life and then got a graduate degree herself.) He seeks out women who, though they may bring no happiness or may even wreak havoc in his life, do not challenge his bond to his mother and his childhood.

What Herzog takes from his family is a tendency to manipulate emotions to avoid deep adult attachments. He uses sentimentality about the idea of family as a tool for avoiding his wives and children. Believing he wants custody of his little girl, he gets excellent advice from his lawyer about how to obtain it. But his intelligence immediately fails. He overhears a child abuse case in court on the way to his lawyer, flies to his little girl, gets a gun and resolves hysterically to protect her. He finds her being lovingly bathed by his former best friend. During the following day, which he spends with her, it is he who involves her in a car accident and gets arrested for possession of the gun. His comic ineptitude and great rush of sentiment win him credit for wanting his child while making it difficult for him to obtain her, and further involve his brothers in his life, for they must rally round to bail him out of jail. This absentminded professor's lack of clarity about himself expresses his exploitativeness.

Herzog's mind moves toward the highest human ideals, even as he behaves badly. While others blame him, he remains enshrined in his intellectual idealism, attacking not his own faults but other people's faulty arguments. In a letter to Erwin Shrodinger on *What Is Life,* Herzog disapprovingly mulls over some of Shrodinger's comments on human nature: "Reluctance to cause pain coupled with the necessity to devour . . . a peculiar human trick is the result, which consists in admitting and denying evils at the same time. To have a human life, and also an inhuman life. In fact, to have everything, to combine all elements with immense ingenuity and greed. To bite, to swallow. At the same time to pity your food. To have sentiment. At the same time to behave brutally." He objects to this omnivorousness as immoral.

But Herzog has life all ways at once by expressing his rage through love. The fabled affectivity of this family man takes the form of manipulating other people's sentiment for his own use. Playing the stumblebum, the wise fool, the total victim, Herzog victimizes others, not only through his dependency but by arousing expectations he will not fulfill. To his wife he offered the life of intellect, but used his superior abilities to wound her; to his new lover he offers his depth of emotion, his genuine sentiment, yet he mocks her even as he shows her off. Even as he enjoys her he remains contemptuous of the extent of her sexual desire. His first wife was too like his mother, his second too like himself. His lover is unreal. But his ability to both involve and frustrate all of them only strengthens his tie to his mother and his sense of himself as the beloved, helpless son who must be pulled and pulled through snowstorms because of the sheer delicacy of his mind.

Bellow's brilliance as a novelist is eclipsed only by his sentimental infatuation with his characters. He subscribes to Herzog's view of himself as "a man who tried to be a marvelous Herzog, a Herzog who, perhaps clumsily, tried to live out marvelous qualities. . . ." Bellow arouses expectations for Herzog which Herzog surely cannot fulfill. At his country house, in which birds have built nests and vines grow through the walls, Herzog lies down, happy, cared for, visited by his reassuring brother, awaiting his lover's sexual ministry. His woman and his brother praise him for accepting the mammoth responsibility of visiting his son by his first wife for one day. Bellow suggests he will work again; this is a new beginning. But it is change by retrogression, by going back. The wash of Bellow's sentiment covers the infantile rigidity of a character who is a prisoner by necessity, who stays locked in himself, lying down happily. He feels too ill-used to live; he will be cared for. Both conditions return him to his greatest sense of security and self-love.

Bellow is an elegiac writer. He celebrates the tragic quality of all human relationships, praising the man whose mind ranges everywhere but over his own character and into the gap between his values and his life. The dark side of family feeling rarely surfaces directly in his novels but is expressed in his play, "The Wrecker," about a man who breaks up his condemned apartment while his wife and mother-in-law ward off irate city officials. Then they take turns wrecking "the home." This is Bellow's most explicit reminder that one of the strongest family feelings is rage.

The answer to violence is submission, is the pull toward the security and peace of passive suffering. Bellow's power as a novelist makes real and arresting the plight of the man who fears his own rage, who pays for his intellectual combativeness by his failure with women, the man who is his own adversary. His suffering is both the beginning and the goal of his life. *Seize the Day,* Bellow's excellent short novel, projects the frustrated dependency of a middle-aged man on his father. Money is the symbol of control and submission, of the powerful father and powerless son. At forty-four Wilky is down to his last seven hundred dollars and besieged with bills from his estranged wife. He begs for money and sympathy from his father, a successful doctor who at eighty can still invite the flattery of everyone in his apartment hotel. Asking for help, Wilky hears his father rehearse his faults, giving the impression of how unfair the father believes it to be for the "better man of the two and the more useful, to leave the world first." Dr. Adler tells him, "I'm as much alive as you or anyone. And I want nobody on my back. Get off! And I give you the same advice, Wilky. Carry nobody on your back."

Wilky clutches at Tamkin, a self-styled financial wizard, to save him economically. But his need for a savior, a father, only makes him an easy mark, and he is quickly swindled out of what little is left. Tamkin asks him,

> "You love your old man?"
>
> Wilhelm grasped at this. "Of course, of course I love him. My father. My mother—" As he said this there was a great pull at the very center of his soul. When a fish strikes the line you feel the live force in your hand. A mysterious being beneath the water, driven by hunger, has taken the hook and rushes away and fights, writhing. Wilhelm never identified what struck within him. It did not reveal itself. It got away.

Wilky is hooked on the idea of father, of being taken care of. As he screams resentfully into the phone to his estranged wife, "Everything comes from me and nothing back again to me."

The concentrated power of *Seize the Day* comes from its complete absorption in the plight of the man who fulfills himself by becoming welded to his failings, frozen in pursuit of a protector. But the only redeemer is death, for Wilky, having lost everything he had, finds himself carried along by a crowd to a funeral where he bursts into tears, weeping for himself and experiencing the relief at the bottom. Wilky and Herzog could not be less alike. Yet both find relief only in letting go.

In Bellow's best novels the psychological suffering of his heroes and their passive resignation to being victims are justified or magnified by historical circumstance. Modern history has taught all of us our powerlessness, and Bellow at his best can put us inside the minds of men who are solidly on the hooks of the Depression and World War II.

Asa Leventhal in *The Victim* is a Depression product whose sense of the gross limitations on his possibilities is bound up with his vision of ruin as economic ruin. Fear of a blacklist hovers over his mind like a force for pure nihilism and negation. He fears such a list records his failings and will stop his future. He feels "disfigured" by the harshness of his birth. He is the son of a "pack rat," a Jewish immigrant fighting for survival, and a mother who died mysteriously in a mental hospital. He both feels enormous anger and the necessity of controlling it. He has a short fuse; he perceives the world as a jungle. He has an innate assertiveness, but it is victimization as a fear, a self-image, a source of security that controls his view of life.

Leventhal's assertiveness is blocked by his sense of dependency on the gentile world he feels demands his docility. He is suspicious of the friendship of the Willistons, an influential couple whose largesse makes him feel like a beggar. "He had often rather helplessly and dumbly put his difficulties in their hands and waited, sat in their parlor or hung on the telephone waiting while his problems were weighed, conscious that he was contributing nothing to their solution, wishing he could withdraw them but powerless to do so. Inevitably there had been times when his calls were unwelcome and the Willistons' patience overdrawn." Into a life of such anxiety and repressed anger Bellow brings a WASP who accuses Leventhal of victimizing and ruining *him* with his combativeness.

Kirby Allbee, one-time friend of the Willistons and a cocktail party provocateur who enjoyed telling anti-Semitic stories in Leventhal's presence, comes out of nowhere, having become a widower, an alcoholic and a bum, to accuse Leventhal of getting him fired years before and ruining his life. Leventhal believes himself innocent, but his ever-ready guilt over his own combativeness binds him to Allbee, who can paralyze him with his dependency. Allbee unmans Leventhal by irrationally accusing him of malicious aggression. Leventhal loses the power to write to his wife because of Allbee's presence in his house. When he sees Allbee following him, he can practically feel him: "The acuteness and intimacy of it astounded him, oppressed and intoxicated him." When Allbee, drawn by Leventhal's "animal hair," runs his fingers through it, "Leventhal found himself caught under his touch and felt incapable of doing anything." The accusation that he has acted aggressively feminizes him.

It is the idea of victimization that is the monkey on Leventhal's back, the loud, hysterical voice crying "My pain is your fault!" that gives him so much trouble. His physical passivity with Allbee is less a homosexual fear than an expression of his social anxiety, of his impotence before his own increasing anger, of the extent to which meeting an adversary only intensifies his docility, his child-like confusion. Leventhal has been conditioned to be controlled by someone who irrationally accuses him of killing with his appetites and assertiveness. Victim and aggressor in this novel are bound in an intense symbiotic relation. Allbee taps Leventhal's guilty repressed anger; Leventhal is hooked by Allbee's authoritative use of the role of victim. In this sense Allbee and the gentile society he represents are peculiar maternal figures, infantilizing and manipulative. Leventhal needs to structure life so as not to see himself as a man of aggressiveness and appetite; he needs to see himself as harmless, even as a loser. He finds in Allbee's pain and failure the spur to his own mysterious sense of failure which exists despite his comfortable job and loving wife. Victimization controls his combativeness and structures his life. He needs his anxiety as a form of security.

Henderson the Rain King and *The Adventures of Augie March* attempt to get away from such insistent fatalism. Henderson is the man of mammoth proportions whose wealth, wives and sheer "strangeness will help to abduct man from life." But Henderson's *joie de vivre* is not believable, and Bellow brings him to a "recognition" that heroism is not simply being without fear or inhibition but is absorbing life's blows and not passing them on to others, not ridding oneself of pain by becoming the aggressor. Henderson has the possibilities of a joyous man, but the emotions of a depressive. He deals with his unexplained guilt by idealizing the man who is a shock absorber. Augie March avoids the pain of involvement by making his life a series of abortive adventures, by maneuvering between the intellectual traps set by "theoreticians" and the worldly traps set by "reality instructors." Bellow keeps his adventurer safely disengaged from responsibility and brings him home, unencumbered with a wife, to his mother.

In Bellow's world, even survivors lose. Mr. Sammler in *Mr. Sammler's Planet* is a seventy-year-old victim of the Holocaust who hid from the Nazis in a cemetery and emerged alive. This Lazarus is an intellectual whose behavior is kind but whose heart is full of contempt for the carnival of violence he sees as the streets of New York, as his daughter's insanity, as his sexy niece's white lipstick which makes him think of perversion, of her "performing fellatio on strangers." His generous nephew who has supported him is dying, his nephew's children are disappointments, renegades. He believes the planet is in decline. His fatalism is intensely conceived and linked to the Holocaust, both as experience and as a model for the destruction of the human family, and of the earth itself.

In Bellow's finest novels, the Jewish man on the hook, or the man whose life is a combat he will lose, is the image of history as personal experience, as tribal death, as intellectual adversary. In *Dangling Man*, Bellow's first novel, the world is all war and death. His Joseph is waiting to be drafted into World War II, losing out in his marriage and connected only through his despair and anxiety to the war-torn outside world. In a stunning dinner-party scene, he perceives his friends as compulsive adversaries who talk not to communicate but to score points. He tries not to get involved, detaching himself by intellectualizing their nastiness. He thinks the purpose of parties is "to free the charge of feeling in the pent heart and that, as animals instinctively sought salt or lime, we, too, flew together at this need, as we had at Eleusis . . . to witness pains and tortures, to give our scorn, hatred and desire temporary liberty and play."

Joseph does not give his rage free play. He plays the observer at destructive parties; he uses his mind as a weapon to neutralize his own combativeness. Split into three voices variously demanding to be heard, to be right, to win, Joseph's mind appears as the Dangler, as the Spirit of Alternatives or as Tu As Raison Aussi. More than merely an expression of his ability to see all sides of all issues at once, this fragmentation reflects the extent to which the obsessional quality of his thought does and undoes itself. His thinking does not lead to any resolutions of his problems but instead provides a model of his defensive techniques and a victory for the problematic. Joseph cannot escape his confusion by force of thought so he tries to avoid it through detachment.

He dreams a dream of historical connections he would like to avoid:

> A few nights ago I found myself in a low chamber with rows of large cribs or wicker bassinets in which the dead of a massacre were lying. I am sure they were victims of a massacre, because my mission was to reclaim one for a particular family. My guide picked up a tag and said "this one was found near . . ." It must have been Constanza. It was either there or in Bucharest that those slain by the Iron Guard were slung from hooks in a slaughterhouse. I have seen the pictures. I looked at the reclining face and murmured that I was not personally acquainted with the deceased. I had merely been asked, as an outsider . . . I did not even know the family well, at which my guide turned, smiling, and I guessed that he meant . . . "it's well to put oneself in the clear in something like this. . . ." As long as I took the part of the humane emissary, no harm would come to me. . . . The bodies . . . were lying in cribs, and looked remarkably infantile, their faces pinched and wounded. I do not remember much more. I can picture only the low-pitched, long room much like some of the rooms in the Industrial Museum in Jackson Park; the childlike bodies with pierced heads and limbs; my guide, brisk as a rat among his charges; an atmosphere of terror such as my father many years ago could conjure for me, describing Gehenna and the damned until I shrieked and begged him to stop.

Jews are clearly victims of the Nazis, clearly facing real and insurmountable odds. The massacred innocents on the hooks could be figures for Joseph himself. Yet he prefers to play the "humane emissary" among them, to avoid both anger and the recognition of himself in the slaughtered victims who remind him of his childhood, of fears of the Jewish hell his father could conjure. To feel connected to these victims is to feel responsible; to feel connected to one's own sense of victimization is to face one's personal responsibility for one's own fate. A prototype of Bellow's later characters, Joseph wants to do neither, preferring to avoid guilt. Like Herzog, Joseph has difficulty assessing his responsibility for his own actions. Because, like Herzog, his thought stops short of himself, he is unable to perceive the extent to which the human emissary is a dangler who derives, by dangling between the guard and the victims, freedom from both. Yet Joseph avoids seeing that he wants this freedom, or that he actively seeks out opportunities in his life with women as well as in the style of his thought, to dangle.

Bellow's characters finally save themselves from nervousness by force of their depression. The anxiety detachment produces is too much even for Joseph to sustain. He resolves when death comes not to think of resisting or of laying "any but ironic, yes, even welcoming hands on his shoulders." He fulfills his own antagonisms by consenting to his own destruction.

By not distinguishing between victimization and vulnerability Bellow is invariably able to put suffering into a political perspective; it is always someone else's fault, never a defect in character. This works beautifully in novels like *Seize the Day* where the hero is a victim of a kind of rigid psychological determinism that seems to control his life. It works well in *The Victim* where Depression concerns and the plight of the Jewish immigrant in America enlarge the hero's personal situation. It works in *The Dangling Man* where the war figures as the cause of personal anguish and combat.

History has given Bellow his best subjects, but history has also taken them away. Depression and war reverberating in the minds of powerless men are large and excellent subjects. But the further Bellow gets from them the more trivial and manipulative the resentments of his heroes seem. Peace and affluence have made the wife, the other woman, the sexual dilemma the adversaries that remain. While Bellow can make a superb entertainment in *Humboldt's Gift* of the foiled love affair, the novel's richness comes finally only from Bellow's capacity for extended intellectual argument. Bellow is making arguments, not novels. Relations between the sexes are his subject but not his forte. Citrine revives himself more through Humboldt's posthumous gift of a manuscript than through his fulsome mistress who leaves him for a man who will marry her, managing to leave this lovable victim in the lurch. It is the gifts of the past that mean most to Citrine and that Bellow himself is blessed with. Bellow is easily our finest elegiac and nostalgic writer. If his heroes use suffering coercively, aggressively, Bellow nevertheless celebrates their unhappiness as the mark of their nobility, and by doing so renews the oldest affirmations of human pain.

Joseph Heller blasted the idea of suffering as the mark of nobility with his jokes. *Catch-22* is the by now legendary novel of World War II which may have set a style for detached humor about any kind of violence or victimization. Through gag after gag, Heller strips war of any meaning but personal greed and sees victims as victims of the army. For Yossarian, the hero, the only purpose in war is to stay alive; for his friend, Milo Minderbinder, it is to get rich. Both see themselves as outsiders. Minderbinder uses his detachment from "the system" to exploit it; Yossarian to get out of it.

All evil in *Catch-22* is external, coming from your own army or your friends. The novel's main conflict is between external strategies: the army's attempt to close all loopholes of escape, the hero's determination to open them up. The army is a model of an American corporation, led by men who "would not even come to an orgy unless [they] could do business there." The power of the book comes from Heller's ability to make all appetites and all suffering seem external, to create not characters but cartoon figures of vulnerability and power.

Through these Heller mocks both power and vulnerability. What wins out is Yossarian's fantasy that there is a world without either, without a "system" of winners and losers.

Apparently the "system" of power relations is everywhere. In *Something Happened* it is Bill Slocum's inner life, the mass of aggressive and destructive impulses he harbors and cultivates while believing himself trapped by his corporation and family. His enemy is what remains of his capacity for caring. Not a large tragedy, *Something Happened* is a sad revelation which may have set out to be Dostoyevskian, a *Notes from the Underground* for the 1970s, but is such a puerile confessional it does not succeed. Repetitious and overly long, it is ungainly, lacking the pacing and style of *Catch-22*. Yet it has the gloss of honesty about a character who, vulnerable himself, knows how to manipulate the vulnerability of others and control them through their need for him.

Bill Slocum is an affluent corporation man whose work is ill defined and insignificant. Although he is acutely competitive at work, he remains obsequious to his superiors and cautious. He is determined not to feel like a loser, and not to suffer. His anger is covert, expressed in the sadistic quality of his friendships and his love, in his need to invent and win arguments with his children. He uses his malice to keep people at a distance, remaining aloof even from his nine-year-old son whom he claims to love most but who is less a real character than a projection of his own fear of vulnerability. He is bound to his family largely through resentment and abuse:

> When I see my wife or children doing something improper, or making a mistake for which I know I will be justified in blaming them, I do not intercede to help or correct, but hold back in joy to watch and wait, as though observing from a distance a wicked scene unfold in some weird dream, actually relishing the opportunity I spy approaching that will enable me to criticize and reprimand them and demand explanations and apologies. It horrifies me; it is something like watching them back fatally toward an open window or the edge of a cliff and offering no warning to save them from injury or death. It is perverse and I try to overcome it. There is this crawling animal flourishing somewhere inside me that I try

to keep hidden and that strives to get out, and I don't know what it is or whom it wishes to destroy. I know it is covered with warts. It might be me; it might also be me that it wishes to destroy me, and I, succeeding in stifling my anger beneath a placid smile, say: Pass me the bread, will you dear?

A compulsive seducer, Slocum vents most of his anger in covert contempt for women:

> . . . they are easy because they are sweet, and they are sweet, I think, because they are dumb. Were it not for the element of status, I really would rather not give orgasms to any of them but my wife, and there's even an element of sadistic cruelty (not consideration, not understanding) in that. Some of them change so grotesquely. They ought to be ashamed. There really is something disillusioning and degenerate, something alarming and obscene, in the gaudy, uncovered, involuntary way they contort. It's difficult not to think lots less of them for a while afterward, sometimes twenty years.

The conscious need to bait women ("I wish these women's lib people would hurry up and liberate themselves and make themselves better companions for sexists like me") makes the confessional more childish than abrasive. The narcissism at the core of Slocum is passive, the regression of a man who knows how to make helplessness work *for* him. "I have something more potent than an ordinary hypocritical male chauvinistic double standard to give me the strength and determination to walk out: I have insecurity. God dammit—I want to be treated like a baby sometimes by my wife and kids. I've got a right. I'm not one of these parents that expect to be taken care of by their children in their old age. I want my children to take care of me now."

The aggressive "victim," the politician of vulnerability, always wants help from the people he hurts. He is a prisoner of his sense of being overburdened. Bellow and Updike, both superb novelists, reflect opposite solutions to a sense of male suffering. In Bellow's novels, the hero's pain insures his lovability and wards off the competitiveness of other men. His willingness to suffer is idealized as a sign of his nobility. Updike's heroes suffer less willingly, using their disappointment as a spur for action, for release. Their actions are circular, but by returning them to their problems also return them to perpetual departure points. The goal of suffering in Bellow's novels is release into a childlike sense of family protection; in Updike's it is a sense of rejuvenation through sex. The determinism of history or of psychology operates to make the suffering described in their novels larger than the ability of any individual man to resist.

Heller offers in Slocum a character who is infatuated with his own emotions, his own sensations. He looks to sex to make him feel alive, but he does not enjoy anything anymore except perhaps thinking about what he missed, the woman he didn't get, the damaged parts of his life. The frustration he wants to escape paradoxically obsesses him. The whining quality of his thought is an antidote, protecting him from acting on his impulses by narcotizing them. When he acts, when he reaches out to love another, he wounds or worse. Trying to comfort his son at the end of the novel, he kills the boy. His airless, ruminative narcissism is a less dangerous alternative than active love.

As a victim the hero finds protection against the harm he would do if he had power. He may wound through coercion, or through frustrating the women who get involved with him, but the pain he inflicts is less than it would be if he consciously and actively sought to give trouble. Slocum at heart is as willing to kill for what he wants as the career officers in *Catch-22* who send their men out on crazy missions in the hope of becoming generals. Herzog and Citrine take a beating to avoid giving one. Rabbit Angstrom in his anger brings about a girl's death. As his wife's numbed, resigned husband he does less harm. Victimization may be the way these men control or manipulate others, but it is also a means of controlling themselves.

The refusal to be consciously, overtly aggressive as a man may have a counterpart in the boyish innocence J. D. Salinger did so much to idealize. *The Catcher in the Rye* compresses a revolt against the life of an adult male into a few days during which Holden Caulfield flunks out of Pencey Prep and comes to New York looking for something or someone to show him the way to grow up. His search for a "father" is shattered when his favorite teacher frightens him off with a homosexual gesture. His romantic love for a girl is ruined by his discovery that she has had some sexual feelings toward a boy and is, he believes, cheapened by this. He tries a prostitute, but is too depressed by her to want her or even to see her as human. Visiting his kid sister's school, he desperately tries to wash all the obscene graffiti off the walls, but he fails. He cannot avoid the sexuality that is all around him but seems missing in himself. His one consolation is his dead brother, Allie, to whom he prays each time he gets to a corner for help crossing the street. Holden rejects all the *rites de passage* that mark initiation into adult life—graduating from school, sexual involvement, faith that he can reach the other side of the street, or cross the line into adulthood. He appeals to his dead brother for what amounts to the perpetuation of his own childhood, a goal that can only be achieved by crossing the line, like Allie, not to manhood but to death.

The pathos of Salinger comes from his belief in both innocence and doom. The Glass family, as their name suggests, are mirrors of each other's special innocence and breakability. What destroys them is adulthood and everything that signifies. In "A Perfect Day for Bananafish," Seymour, the idol of the Glass stories, goes on his honeymoon, calls his lush wife "Miss Spiritual Tramp of 1948," and walks out to kill himself. Before he does, he makes up a beautiful story about bananafish for a little girl. Will this child grow up to be "Miss Spiritual Tramp of 1958" and spend her time polishing her nails and talking on the phone? In "Uncle Wiggly in Connecticut," her future may be suggested in the life of the Connecticut woman married to a conventional bore. She adores her dead lover Walt, whose specialness is often on her mind, and whose loss has obviously left her life empty. Salinger's gentle, brilliant boys cannot grow up: if they do, they die. If they marry, they can keep their specialness only by suicide. In death they remain ideals of tenderness, sweetness, unworldiness. The closeness of the children of the Glass family—the parents are not particularly visible—reflects the faith that people can make a connection as brothers and sisters more easily than as husbands and wives. Nostalgia for the family one had as a boy is an antidote to the sense of responsibility and effort involved in the family one makes as a man.

Holden Caulfield and Seymour Glass are undefended against the facts of the world. But you might say their defenselessness is their defense against the usual expectations the husband and father has to meet. Their inability to face harsh facts protects them from being expected to face them, even though it leaves them only death or regression as alternatives. The hero as an innocent boy, as a man who prefers death to his bride, may be an exaggerated version of the hero who must remain apart from conscious, active competition with other men, and who ambivalently offers his aggression as a sacrifice to his everlasting, boyish charm.

The hero as victim opens up the plight of the adult male character in a feminized world. The middle-class suburbs of Updike's later novels, the preoccupations of Bellow's Citrine or of Herzog, are curiously without the usual male obsession with work or sports. Slocum approves of his son's hatred for the gym coach who wants to instill competitive values in him. Jerry's interest in work is sated by an even money flow. He competes only in mixed volleyball. St. George had his dragon, Sir Galahad his Grail, but many new heroes have only women to structure their lives and define them as men. These prisoners of sex never really revolt. They retreat from their anger, they try not to inflict the degree of pain they know they can, but they wound through arousing and frustrating love. They are heroic not in a traditional sense, but in the magnitude of their submission to what they see as the unalterable facts of a world of women.

VI

AMERICAN REBELS ARE MEN OF ACTION

Every ethic based on solitude implies the exercise of power.

Albert Camus, *The Rebel*

I have no doubt, however, that there had been moments during the writing of the book when I was an extreme revolutionist. I won't say more convinced than they but certainly cherishing a more concentrated purpose than any of them had ever done in the whole course of his life. I don't say this to boast. I was simply attending to my business. . . . I could not have done otherwise. It would have bored me too much to make-believe.

Joseph Conrad,
"Author's Note" to
The Secret Agent

Rebellion in recent American fiction has not involved a direct attack on the system so much as a parody of its values. The rebel and tyrant may hate each other. But in our holistic fiction of aggression they share a common preoccupation with power and a tendency to see life in terms of winning and losing. They equate, in the novel, political goals with one-upmanship.

One of our best rebel writers, Norman Mailer, has always wanted to take a political stance in his novels. Although his opinions may change, his stance is always the same: the perpetual adversary whose character seems an argument against every other character. In *The Naked and the Dead* (1948) he took on the establishment that waged the "good" war. In *The Deer Park* (1955) he attacked those who attacked the establishment in the 1950s by refusing to testify before the House Un-American Activities Committee. In *An American Dream* (1965), a lesser novel set in the sixties, he played iconoclast to the sexual revolution. Mailer has avoided the traps of every kind of political orthodoxy by being against everybody *else's* power trip.

Mailer has explored power as political fact, as sexual energy, as narcissistic joy. In *The Naked and the Dead* he wrote of the army as the model of the future, the prototype of all social organizations run according to rank and fueled by fear and aggression. His General Cummings is not exactly against fascism; he is a dictator himself with a great capacity for calculated cruelty. He insists that the ability of men to kill the enemy is determined by how much they hate their officers and can displace their frustration on the officially designated target. Cummings judges a man's worth by his readiness for combat, by the swiftness with which he translates his anger into an act. "If you're holding a gun and you shoot a defenseless man, then you're a poor creature, a dastardly person. That's a perfectly ridiculous idea, you realize. The fact that you're holding the gun and the other man is not is no accident. It's a product of everything you've achieved: it assumes that if you're aware enough, you have the gun when you need it."

Lieutenant Hearn, a Harvard man who is Cummings' orderly, is the rebel against so much masculine aggression. He argues that the war will be won not by hate nor even material superiority but by the persistence of ethical ideas. Yet he is himself driven and drawn by his aggression. He is simply too unable to break through the sentimentality and guilt he feels over his own combativeness to acknowledge it. He is uncomfortable seeing Cummings as anything but the protofascist general. When Cummings tries to share the vulnerable side of himself, Hearn is revolted. Pained, Cummings confesses, "My wife is a bitch." Hearn mumbles, "I'm sorry," in such a way that he both magnifies and rejects Cummings' admission of his humiliation at his wife's hands. Hearn is both sentimental and unfeeling, full of anger and incapable of empathy while professing to value compassion. He pays for his ambivalence with his life.

Instantaneous vindictiveness releases Cummings from self-hate and embarrassment over his admission. He sets in motion a set of circumstances that reveals Hearn's delusions about himself, his idea that life between men, between those who have power and those who do not, can be anything but hate and combat. Hearn, discovering he loves power, loves leading his platoon of men, is so blinded by the myth of the beloved chieftain that he overlooks the resentment of the sergeant he outranks. He trusts him and his men and is led into an ambush by them.

Mailer's marvelous cast of characters are men made obsolete by the Depression and rejuvenated by the army, men who have no place in the outside world but have found channels for their resentments in war. The instrument of Cumming's wrath is Croft, a redneck fom Texas who learns at an early age that the man who does not shoot fast gets humiliated. Tracking a deer in Texas as a boy, he carefully gets the deer in his sight, but hesitates, "trembling before he can shoot." His father shoots the deer and jeers at his son. "The tears freeze on the boy's eyes and wither. He is thinking that if he hadn't trembled he would have shot the deer first." When Croft meets Hearn, who has been assigned to lead his platoon, he assesses him as a man bewitched by authority, but too weak to assume control. He resolves to eliminate him and he does.

War is a metaphor for the world as combat. War has no ideological value in this novel but offers a chance for the Cummingses and the Crofts to put their personal stamp on reality. The jungle warfare on the imaginary island of Anopopei slides into a sense of the Darwinian jungle from which the fittest for survival emerge. The purpose of Mailer's fittest characters is "to achieve God. When we come kicking into the world we are God, the universe is the limit of our senses. And when we get older, when we discover that the universe is not us, it's the deepest trauma of our existence." Croft parodies in his sexual life the same will to subjugate. In sex he is violent, his mind screams, "Crack that whip! I hate everything that is not myself!" Power in sex and war is a Faustian trade-off of sentiment for the illusion of immortality and control, the illusion that the universe is you.

If Cummings and Croft are projections of Mailer's ambitions, Hearn embodies his anxieties about himself. Mailer at twenty-five emerged as the adversary of both Cummings and Hearn by giving victory in war neither to the ideologue nor the fascist, but retaining it for himself, the novelist who manipulated the jungle, the general, and the ideologue. He gave the conquest of the island of Anopopei to chance and to nature, which kills the Japanese with jungle diseases. "What is a novelist," Mailer would ask in the *Prisoner of Sex,* "but a general who sends his troops across paper?" The novelist as general is concerned with character as strategy, as the alternation between a hunger for power and a fearful recognition of ineffectuality.

Mailer's novels are products of his ambition and frustration. As *The Naked and the Dead* was obsessed with power, so *The Deer Park* is obsessed with powerlessness. As Hemingway presented the war of man against his own insignificance as a war against nature (the big fish, the river, physiological impotence), Mailer presents that war as a sexual and social battle in which politics, art and sex are aspects of the same condition. The consolidation of power in the 1950s and the translation of combative, competitive values into personal relations are handled through the politics and loves of Charles Eitel and Sergius O'Shaugnessy. Charles Eitel refuses to testify before HUAC, although he has no political convictions in common with or no personal affection for the people he is asked to testify against. Nor does he know much that would harm them. For his refusal he is blacklisted, stripped of his home, which he can no longer afford to maintain, and forced to live in a vacation home in the desert which he had bought in better days. Repression is not only a political fact in the novel, but also a metaphor for Eitel's emotional life.

Eitel is an opportunist, a maker of successful commerical movies who, "to find his self-respect. . . . usually had to do something which was of no advantage for himself." He sees his noble gesture with only irony and contempt. He makes fun of the other holdouts against the committee; he sees himself as a loser and is contemptuous of anyone who is willing to get involved with him. No lovely, young, desirable, ambitious woman would have an affair with this man of fallen fortunes but only those "so low in the scale of Desert D'Or that he would still be important to them." A man's desirability is directly measurable in this novel by his wealth and influence. Without either, Eitel is impotent. What is curious in this novel is not that Mailer sees marketplace values as the ones by which women judge men, but as the values by which a man judges himself.

Mailer is a master of the play of power and obligation between men. When an executive at Eitel's old studio, Collie, comes to Eitel for a favor, asking if Eitel will care for a mistress he must regretfully get rid of, Eitel agrees, glad to have Collie obligated to him. He uses Elena to settle scores with Collie's boss, but finds himself drawn by her vulnerability, her near-beauty, her pride and her sexual gifts, which are so great he believes an affair with her could "return his energy, flesh his courage and make him the man he had once believed himself to be." But in regaining his sense of himself as a man, he only regains his need to humiliate and wound.

The vision of the world as unrelieved aggression, appropriate in a war novel, is translated in *The Deer Park* into a vision of male combativeness toward women. Love inspires not protectiveness but anxiety and aggression in the man who feels that "his delicate manhood depended on Elena." The more he needs her, the angrier he becomes, the more demanding of her and the less giving. While he becomes abusive and cold, "what he could not bear was the thought that she did not love him completely without thought or interest in anything else alive." He resolves to get rid of her when it will not be "too disturbing to his work."

> The proper time was in a month, two months, whenever he was finished; and in the meantime adroitly, like fighting big fish on slender tackle, he must slowly exhaust her love, depress her hope, and make the end as painless as the blow of club on the fatigued fishbrain. . . . He was cool as any good fisherman. "I'm the coolest man I know," he would think. With confidence, aloofness, and professional disinterest he maneuvered Elena, he brought her closer to the boat. . . . He could not let her realize how his attitude had changed; she would force a fight which would go too far; that was her pride; she would not stay a moment once she knew he did not love her. . . .
>
> He had wrapped his work about him and it gave the distance he needed, the coldness, the lack of shame. He would be far away

from her, he would eat a meal without speaking, his eyes on a book. He would sense how despair swelled in her fatiguing love, fatiguing spirit, and at the moment when he would feel that she could stand it no longer, "we can't go on like this" about to burst from her mouth, he would confuse her completely.

"I love you, darling," he would say out of a silence and kiss her and know her bewilderment had seated the hook more firmly.

Eitel is guilt-ridden over what he does, but his guilt is a spur to self-interest. He offers to marry Elena when things go bad between them, and to get divorced. "I mean I know how much you'd like to get married because you feel that no one cares about you that much, and I want to show you that I do." Marriage would convert the dependency and attraction and guilt he feels toward her into an external obligation he could meet and discharge. "If they didn't marry, he would remain wedded to her," he thinks fearfully.

Sergius, his foil, is the artist as a young man on the rise whose success lies in the possession of Eitel's desirable ex-wife, the movie star Lulu. In bed with her, he has, he says happily, "a million men riding on his shoulders, her fans." Unlike Eitel who now cannot bear to hear of Elenas's old lovers, Sergius "was charitable to all of Lulu's. She had sworn they were poor sticks to her Sugar. I wanted to set Eitel at my feet, second to the champion. It pleased me that in my big affair I had such a feel for the ring." Lulu gives him the sense of being a winner not only in being seen with her, in being in bed with her, but in the roles she assigns him in sexual fantasy—such as the guy who persuades the teenage virgin to say yes. Sergius derives his sense of self from her; he is the young man attached to the famous woman. He becomes acquiescent through his need for her status.

There are no female heroines in Mailer's novels because women are the territory to be conquered, the mark of a man's ability to compete with other men, displace other men from a woman's mind. The sexual world of Mailer's novels tends to divide between rich, prominent women a man needs for social gains, and submissive women who make him feel strong. The former finally exact the price of subordinating their men and making them feel like failures. Lulu gets rid of Sergius to marry a homosexual actor she does not even like but whose fame will improve her career. Elena leaves Eitel, but after a severe auto accident she calls him, and he marries her out of guilt. However, his fortunes restored, his career in excellent shape, he accepts her as his cross to bear and has an affair with his own ex-wife, the ever-popular Lulu. This portait of the artist in middle age is a lesson in how not to be vulnerable to any of the women in your life.

Mailer's men balance between women who abuse them and women they abuse. They are prisoners not of sex but of life as a pecking order. They seek authority by enlarging themselves by means of submissive women, half-hoping to be made tender by them, or they try to fulfill their dependency and relieve a fear of failure through allying themselves with powerful, ruthless women. They cannot quite accept their own ruthlessness nor are they able to be as loving as they would like to be to their sweet, lowly ladies. They go from one kind of frustration to another, to such an extent that their life is shaped by anger, and they become sexual guerillas.

"Living with her I was murderous; attempting to separate, suicide came to me," says Rojack in *An American Dream* (the dream is presumably to murder one's wife and have sex with her maid). Rojack's wife is a "Great Bitch," a rich woman who binds him to her through the sustained sense of anger and impotence she arouses in him.

> Deborah was an artist with the needle and never pinked you twice in the same spot. . . . So I hated her, yes indeed I did, but my hatred was a hate which wired my love, and I did not know if I had the force to find my way free. Marriage to her was the armature of my ego; remove the armature and I might topple like clay. When I was altogether depressed by myself it seemed as if she were the only achievement to which I could point. . . . I had also the secret ambition to return to politics. I had the idea of running someday for Senator, an operation which would not be possible without the vast connections of Deborah's clan. . . .

The other side of Rojack's ambition is death. Rojack sometimes feels his personality was formed on the void. Since the war he has felt lost in a "private kaleidoscope of death," ushered in by a dying German soldier whose "eyes had come to what was waiting on the other side, and they told me then that death was a creation far more dangerous than life. I could have had a career in politics if only I had been able to think that death was zero, death was everyone's emptiness. But I knew it was not." Death is an active principle, the anger that motivates Rojak's sexual energy.

Mailer's men need to see sex as an aggression. (If it is not, as it is not initially between Eitel and Elena, they need to covert it into a combat). Rojack's destructiveness takes the form of wanting to steal something from women. From a prominent woman, he wants status and privilege. From the have-not, he wants the greater gift, rebirth. After killing Deborah, Rojack meets a pretty blonde nightclub singer, Cherry. Making love to her, he resents her diaphragm. "I searched for that corporate rubber obstruction I detested so much, found it with a finger, pulled it forth, flipped it away from the bed." He wants her to conceive and believes she does. "There was a child in her, and death, my death, my violent death would give some better heart to that embryo just created, that indeed I might even be created again, free of my past." Fathering the child means being the child. Birth contains the promise of a new personality. Like Eitel with Elena, Rojack needs to be reborn through Cherry into a world without aggression where he is able to feel tenderness.

Mailer's men deal with their attraction to women by assaulting not the women so much as their own desire. Before he beds and de-diaphragms Cherry, Rojack rakes her over for faults, becoming hypercritical of her laugh and feeling that "a tension had begun in me that she be perfect." As Eitel emotionally bludgeons the all-loving Elena into despair and flight, so Rojack, by loving Cherry, puts her in a position of danger. She is killed by another man while Rojack is away.

Mailer's men know they will have to change to be happy,

will have to kill their anger to be reborn. While Rojack may call the diaphragm a corporate obstruction, the industrial impact on his life isn't birth control but dependency control, armor against giving in. The anger of these men becomes a defense against letting go, against passivity; through it they stay in competition with other men. Women are the instruments of that competition. While men are not brothers to each other, they are in Mailer's novels brothers in their feelings toward women. Having murdered his wife, Rojack is absolved by his father-in-law, who admits a long incestuous affair with his daughter and acknowledges he wrecked her life, too. They part friends. Rojack cries with the cop who has just beaten his own wife over the beaten-to-death body of Cherry. Live women offer a challenge to a man's power, but dead ones provide an opportunity for men to console each other. It is only with the deaths of both the rich, withholding women and the sexy waif who gives her all that the hero is free.

Anger, and the adversary relation, remain the fixed principles in Mailer's work. *The Naked and the Dead* and *The Deer Park* stated the problem in terms of the concerns of the forties and the fifties; *An American Dream,* though inferior to both of these fine novels, gives an image of the changed circumstances of the war between the sexes. Rojack is in "the void," cultivating his anger as a way of dispelling emptiness, as a way of avoiding his own vulnerability and passivity. Without the political ramifications of Mailer's earlier novels, this one reaches into the heart of sexual politics. There flourishes male fear of not performing, not succeeding, not being able to feel alive. To remain with the Great Bitch is to know one kind of abdication; to be reborn through love is to dissolve in another infancy. Fear of both possibilities keeps Mailer's men running from one to the other on the treadmill of their own anger. Their anxiety is an antidote for their passivity.

Mailer's men make dislocation the stable principle in their lives. Like other, more famous male leads (Bogart, Cooper), they play one role forever and live the myth of endless adolescence. Disillusionment with one woman catalyzes the lust for another. What they revere is not manhood, but the promise of manhood. This promise is a screen against seeing how little they want to fulfill it.

What do these men want? Mailer's last unchallenged value is energy in the moment of crisis. His men want to keep the juices flowing against the antagonists that are much more important than women: age, exhaustion and failure. As Mailer wrote in *Advertisements for Myself*, "Movement is always to be preferred to inaction. In motion a man has a chance, his body is warm, his instincts are quick, and when the crisis comes, whether of love or violence, he can make it, he can win, he can release a little more energy for himself since he hates himself a little less. . . ."

In his famous essay "The White Negro" (1957), Mailer wrote that the "Psychopath may indeed be the perverted and dangerous front runner of a new kind of personality who could become the central expression of human nature before the twentieth century is over." Despite his reservations, Mailer praises him as the forerunner of the hipster who has special gifts of self-expression because he is incapable of delaying or repressing his desires. He can

> replace a negative empty fear with an outward action, even if the fear is of himself and the action is to murder. The Psychopath murders—if he has the courage—out of the necessity to purge his violence. (It can of course be suggested that it takes little courage for too strong eighteen-year-old hoodlums, let us say, to beat in the brains of a candy-store keeper, still, courage of a sort is necessary, for one murders not only a weak, fifty-year-old man but an institution as well, one violates private property, one enters into a new relation with the police and introduces a dangerous element into one's life.

In the adversary relation to the police, the wife, legal system, or whatever, Mailer's heroes become gross parodies of the system they attack. As Hearn plays Cummings ineffectually to his platoon, as Eitel suffocates Elena after the committee has taught him what repression means, so Mailer's heroes carry to extremes the self-seeking individualism of the society they attack. They are not psychopaths without a cause, but men for whom the cause is themselves, and anxiety and sexual rebellion are fuels.

Combativeness in Mailer's novels reflects the hero's belief that he can still win against the forces of exhaustion and emotional poverty. From the Depression characters of *Naked and the Dead* to the middle-class have-nots in *Deer Park* and *American Dream*, Mailer's men see power as both an anti-depressant and a real possibility. Robert Stone is a brilliant young writer who sees the world as unrelieved combat in which the best the renegade can hope for is survival. Extreme in its nihilism, *Hall of Mirrors* (1964) is a stunning novel of the hero who replaced the hipster: the man who gets drunk or stoned not to intensify his awareness but to render himself stolid and insensible. He wins freedom from caring. In his America all politics is propaganda; there is no social philosophy that is not reducible to a tool for manipulation. In Mailer's novels the hero wants success on his own terms. In Stone's, Reinhardt, the hero, wants invulnerability.

Rheinhardt is the new man of action whose goal is to defeat his ambition and his love. As a clarinetist he had won the admiration of a man at the heart of the musical establishment by playing Mozart brilliantly.

> There was perfection in this music, something of God in this music, a divine thing in it—and the hungry coiled apparatus in Rheinhardt was hounding it down with a deadly instinct, finding it again and again. . . . So it turned out that morning that just above the barrier of form was a world of sunlight in which he could soar and caper with an eagle's freedom, rule and dispense passion, where his breadth was the instrument of infinite invention. . . . At the end, in the final passage—he could feel himself—the brain, mouth, diaphragm, lungs and fingers of the musician Rheinhardt fused together in a terrible invincible unity. . . .

Rheinhardt for no stated reason gives up the clarinet and becomes an alcoholic. Is it fear of playing a false note, of losing at what he cares about, that has made him a drifter whose wind and mouth are gone? Stone does not say. But his novel offers a vision of America cast in the mold of Southern boss politics that suggest he sees the very desire to hunt, find and become God, as man or musician, as evil. To care that much is to feel ambition so concentrated it exhausts everything else. The musician who feels like God is the benign aspect of a temperament Stone hates but cannot escape. Rheinhardt does not attack the "boss" temperament but adopts it, exploits it and parodies it. He sees that the way to defeat the system is not to develop opposite values but to beat it at its own game.

Rheinhardt, having drunk up all his money in New Orleans, gets a job with a right-wing racist radio station. He exploits their desire for a man who can discover in the news a pattern that reveals the red menace, the destructive impact of blacks and the danger caused by Northerners and liberals. His bosses are "sub-human cruds," remarks one of Rheinhardt's pothead neighbors. " 'You talk like an extremist,' Rheinhardt said. 'You're not seeing the Big Picture. . . . Speaking as a broadcaster it's my opinion that there is a deep confusion in the popular heart and mind. The pop heart and mind demand assurance. Unusual times demand unusual hustles. The explanation number is very big. . . . My conscience is clear,' Rheinhardt said, 'it's bone dry.' "

The cast of drop-outs from the system form a society which is the mirror-image of the larger pecking order. Rheinhardt is brutally rude to a well-intentioned civil-rights activist who protests his racist broadcasts. He is close to no one but closest to Farley, an actor who played Jesus in a passion play and discovered that salvation can be big business. Through Rheinhardt he becomes a preacher on the air. Around them exists a world polarized between the political boss of New Orleans and the swarm of tarred, burned, starving, robbed, exploited blacks and women who are projections of irremediable vulnerability. The civil-rights activist had had nervous breakdowns and seems to imply that people cannot change injustice, they are destroyed by it.

Relations between the sexes only extend antagonisms. Geraldine, Rheinhardt's woman, is from West Virginia, a pretty, uneducated Appalachian whose husband has been murdered, whose baby has died and whose face has been badly scarred by a man who got mad at her. She adores Rheinhardt because he seems to fly so high, so far from his feelings. He is also far from feeling too much for her. She says: " 'You're so wild and you don't have nothing to do with anythin'. . . . I need you, love, . . . I really do.' 'You must be out of your mind,' Rheinhardt told her. 'I don't say things like that to you, why do you say them to me? Man that's an obscenity. . . . If somebody ever tells you, Geraldine, that they need you you tell them to buy a dog.' "

Where Mailer's characters operate with total seriousness to dominate women, Stone's Rheinhardt is ironically self-conscious about his desire to control Geraldine: "First thing, we have to consider my needs. We have to consider them from every possible angle in every minute detail and we have to work tirelessly to gratify them all. That's going to take so much time and we'll be so busy that we won't even have to think about your needs at all." Yet the more Geraldine's body and presence calm him the more despairing he becomes and the more determined not to get too involved with her even though they live together. He has, presumably, tried love and found it wanting.

Politics like sex maps the way life divides winners and losers. The novel's climax is a giant rally sponsored by the right-wing racist establishment and boasting a variety of military and Bible Belt religious groups. Rheinhardt gets stoned as the speeches begin and by the time everyone realizes the rally has been rigged to turn into a race riot which will justify the local boss's vendetta against blacks, all he can think of is the extraordinary greenness of the grass. But he is called upon to save the day, to be God. Asked to calm the crowd, he climbs to the podium feeling as though he will play or conduct. In a hilarious parody of his exaltation over Mozart, he speaks: "Americans our shoulders are broad and sweaty but our breath is sweet. When your American soldier fighting today drops a Napalm bomb on a cluster of gibbering chinks it's a bomb with a heart. In the heart of that bomb, mysteriously but truly present, is a fat old lady on her way to see the

World's Fair. This lady is as innocent as she is fat and motherly. This lady is our nation's strength. This lady's innocence if fully unleashed could defoliate every forest in the torrid zone. This lady is a whip to niggers." The riot continues, heated; the man without convictions still feels happy.

Rheinhardt can walk away from violence, can parody it. Geraldine, trying to reach him, runs from two men she believes will attack her and gets arrested by police who see she is stoned and has a gun in her bag. She hangs herself in jail. Geraldine is engulfed by the anger all around her, like the civil-rights activist, who gets himself blown up trying to prevent someone from dynamiting everyone in the stadium. Lovers and moralists are the first to go.

Stone's message is, "No help." These are Rheinhardt's words to Geraldine when identifying her body; he leaves saying the phrase like a litany. "I'm alive baby. . . . It was you who died. Not me. I don't need you, could you think that. You know . . . I mean . . . it was no great passion, Geraldine. It's me that is going to have the next drink not you. That's what No Help means. Once more . . . One more time. I'm a survivor. I love you baby—No Help."

In Stone's novels pain is a separate and separating experience. Rheinhardt decides to go to Denver and climb the highest hill to look down on the human scene and get as far away from it as he can. His binges, his highs are diversions from the human fight and not an intensification of any emotional experience. This is the hipster, the dropout, the renegade not as an adversary of the system but as its product. He does not escape its ruthlessness, he merely inverts it. This is the hero as the man who still believes that "whenever somebody says something that's a drag you should always say something that's a bigger drag." This is the hero who by choice and will has nothing left but the need to be above it all.

The rebel and tyrant are mirror images of each other. The bitterest possible expression of their connection is at the end of *Dog Soldiers,* (1975) where Hicks, the career marine turned pusher, lies dying in the desert, his heroin stolen by a crooked cop who wants to sell it himself. Hick's last thoughts are:

> You know what's out there? Every goddam race of shit jerking each other off. Mom and Dad and Buddy and Sis, two hundred million rat-hearted cocksuckers in enormous cars. Rabbits and fish. They're mean and stupid and greedy. They'll fuck you for laughs. They want you dead. If you're no better than them, you might as well take gas. If you can't get your own off them, then don't stand there and let them spit on you, don't give them the satisfaction.

Stone's renegades have as a goal not change but exploitation and vindictiveness. Unlike the oppressed characters in Depression novels, Stone's are welded to a sense of the destructiveness even of victims. The new rebel leader is often as afraid of his "constituency," or ought to be, as he is of his establishment oppressor. Ken Kesey's *One Flew Over the Cuckoo's Nest* (1962) is a vision of America as an insane asylum run by Big Nurse who serves the "Combine." Like Stone, Kesey suggests that the country is run by a corrupt force using all institutions to consolidate its power and manipulating all inhabitants, punishing with humiliation or shock treatment or lobotomy. Oppressive society is projected in the novel as Big Nurse, who controls and infantilizes in the name of the best interests of the inmates.

The apparent hero of the novel is McMurphy, who comes to the asylum voluntarily, thinking it will be better there than in the county jail to which he has been sentenced for six months. But the actual hero is Chief Bram, a virtually catatonic Indian whose consciousness forms the basis of Kesey's best writing. McMurphy is a larger-than-life figure whose ebullience and affection for the inmates bring him into conflict with Big Nurse. "We couldn't stop him," says one of the inmates, "because we were the ones making him do it. It wasn't the nurse that was forcing him, it was our need that was . . . pushing him up, rising and standing like one of those motion picture zombies obeying orders beamed at him from forty masters."

McMurphy is pushed by the need of the inmates toward the attack on Big Nurse that gets him lobotomized. Chief Bram seems to have required the sacrifice of McMurphy's energy to give him life. He kills McMurphy, presumably out of love, incorporates his strength and escapes. The model hero here is not the man who dies for the cause, but the man who waits for another to die and is thereby released. Chief Bram runs out, self-propelled, his own man because he has no living attachments, and no capacity for them.

Ironically, it was Kesey's success within the system (he was a Woodrow Wilson Fellow at Stanford; his first novel was a success) that financed a revolt against establishment values that itself turned out to be a parody of authoritarianism. Kesey founded a tribal society, the Merry Pranksters, which became the subject of Tom Wolfe's *The Electric Kool-Aid Acid Test.* LSD was the spiritual center of the Pranksters, but it was dispensed by Kesey, who financed all the group's activities and was called "the Chief." The commune was callous toward people who had too many bad trips and fascinated with the Hell's Angels as a violent "out" group. Despite the buoyancy in Wolfe's depiction of the Pranksters' pranks, his most memorable portrait of Kesey is as the Chief bereft of his tribe, the man who has lost his followers and is hiding from the law in Mexico. This underground renegade comes across as a small-town Napoleon on a Central American Elba.

The rebel does not act to resolve a problem so much as to release his tension or to extend his influence. This unabashed narcissist is not necessarily new. He may even be an old American type. He may even be a Vermont Yankee. John Gardner is one of the few American novelists who has remained fascinated by the man who acts, who wills one thing intensely enough to get it. A medievalist, Gardner seems to have kept the old faith he expressed in *Grendel* when he wrote that "except in the life of a hero, the whole world's meaningless." *October Light* is a stunningly written tragicomic novel that searches through the operative myths of national greatness for a surviving American heroism. Sally, an eighty-three-year-old feminist, and James, her seventy-year-old farmer brother, mirror two heroic strains in the American consciousness, the one intoxicated by progress, the other devoted to the land and deriving its values from the endless repetitions of nature. Belief in progress and change and willingness to work the land opened up the American wilderness. But what happened to these impulses under the press of the civilization they produced?

Gardner's hero and heroine are old-time individualists caught in the world they made. James Page is the Anglo-Saxon hero as a Vermont Yankee. His life with his widowed sister is this concentrated faith:

> He knew the world was dark and dangerous. Blame it on the weather. "Most people believe," he liked to say, "that any problem in the world can be solved if you know enough; most Vermonters know better." He'd had one son killed by a fall from the barn roof, another—his first born and chief disappointment—by suicide. He'd lost, not long after that, his wife . . . he was better than most men at taking [death] in stride . . . he understood what with stony-faced wit he called "life's gravity," understood the importance of admitting it, confronting it head on, with the eyes locked open and spectacles in place. . . . All life—man, animal, bird or flower—is a brief and hopeless struggle against the pull of the earth.

Sally rebels against the pull of her fate, her age, her sex. This WASP as Yankee princess has never had children, has gone traveling with her generous, quiet husband, tried her hand at business and become a Democrat in Republican Vermont. But she feels cheated, resenting her lack of sexual opportunities, her husband's silences and the weight of everything she didn't do. She sympathizes with every underdog, every kind of trendiness and every liberal cause. She is in love with the action-packed world that shines from her color TV. Finding her *in flagrante delicto* with its values, James shoots the TV through its glassy face. When she defends the Equal Rights Amendment, James is furious and chases her up to her room with a stick. There she stays to assert her inalienable right to free speech.

"The hero is he who is immovably centred," wrote Emerson. Gardner's heroes are intractable curmudgeons. James rigs up a gun in a web of string to fire on Sally if she leaves her room. Sally rigs up an apple crate to fall on James if he comes into it. Heroic in their determination not to lose to each other, they are willing to fight to the finish. Who is right? It doesn't matter. Gardner's hero-worship is not for the legitimacy of anyone's cause. The point of the heroic à la Gardner is not to stand for what is right, but simply to stand in a condition of obsession, fully accepting one's own outrageous wrath.

Sally and James caricature the plight of peole who have heroic angers and spites but can find no arena big enough to legitimize their energy. The revolutionary values that created America or the determination that developed the land are no longer in the service of producing a new civilization. Gardner's people are stuck in a world without new lands to conquer in which progress has become a set of small differences, bigger refrigerators, better TV pictures. In frustration, James sentimentalizes the past and simplifies his life to his fatalism. Sally is nostalgic for a future that does not exist. Both find the present a hard time for heroes.

America has been in a phase so inimical to conviction that people who hold on to values often feel ridiculous. The corporation man's etiquette of other-direction has become an ethical code of compliance that has replaced steadfastness as a value. The person who maintains his ground has not kept up with the times. Sally knows everyone thinks she is a fool for not making a false apology to James to regain her freedom. Locked in her room reading a thriller, *The Smugglers of Black Soul's Rock,* even she has fantasies of capitulation which are perversely reflected in the novel she reads. Intertwining with *October Light,* the thriller holds a distorting, funhouse mirror up to it. It reverses the relation of the reader to art. Once the reader whose life was anger and compromise looked for fictional heroes great and firm enough to take him out of his smallness. Sally, whose flintiness is on the grand scale, finds herself consuming a fiction of marijuana smugglers, of accommodators whose only value is self-interest. America supplies us with fantasies which reinforce cynicism. The popular fantasy is of letting go, sinking from hard confrontations into the low tensions of antagonistic cooperation.

Why can't Gardner's heroes make their lives seem worthwhile or transmit their capacity for commitment? Gardner's novels are genealogies in which the hero, by wrecking his children, makes his values obsolete. Sacrificing them to his own ambition, he puts a curse on his own line by the sheer spillover of his aggression. The Oedipal myth comes alive in Gardner's work as a murderous conflict of generations in which sons are never up to the fight. In *Nickel Mountain* Willard, the nasty narcissist, feels his father "nicked him in the balls" and vents his anger by seducing and betraying a loving young girl. In *The Sunlight Dialogues,* the Sunlight Man cannot face his anger towards his father and father-in-law, who mistreat him. He strangles his father-in-law, but then drops into denial of his fury and paints the word LOVE on a four-lane highway. In *October Light,* James's son Richard has been ridiculed by his father all his life, but can only play-act his rage. Dressing in costume on Halloween at twenty, he terrifies his uncle and triggers a fatal heart attack. Consumed by guilt, he hangs himself. Gardner's fictional sons displace their anger, cover it, or turn it against themselves. They are a generation of passives who cannot equal the instantaneous action of their fathers. James Page has no trouble retaliating against his suicidal son. Right after the boy hangs himself, James burns down the boy's house, pursuing him with his anger beyond the grave. Gardner's young men have no exits from their parents' wrath. They try to erase themselves with drugs or alcohol, or are eaten up by guilt. But nowhere can they escape the image of themselves as damaged men.

For Carlyle, hero-worship was the proper attitude toward the aristocracy of genius from which one could draw idols. For Emerson, heroes formed an aristrocracy of purpose. For Gardner, heroes are unloving people caught in, but never diverted by, other people's needs or escape systems. All around them are people who have failed at conviction, lost purpose, failed even at anger. This is the America that believes in nothing and accommodates to everything. In this America the man of conviction is an oddball too often forced to fight for small stakes.

Were heroes better off in the past? At the beginning of historical time even sexual wars were young and huge. In his awesomely conceived 354-page blank-verse poem *Jason and*

Medeia, Gardner offers a Jason who is a calculating manipulator of people. His Medeia is a prototype of female power, a sorceress whose strength comes from the irrational forces of night. Medeia's witchcraft has brought Jason the Golden Fleece, killed her brother and Jason's throne-usurping uncle. Jason, eager to marry a young princess who can bring him a throne, calls his betrayal of Medeia an act of reason. She teaches him the price of his ambition by killing their sons and his new bride. This matched pair is no greater in moral stature than Gardner's vindictive, flinty Vermonters. Jason is a lesser hero who has needed a woman to bring him a prize and used her as a death weapon. Medeia is so consumed by her obsession with Jason that she sacrifices her children to her war with him. What happens between them is not different in kind but only in scale from the conflicts of Gardner's modern characters. Spilling out of the kitchen, the bedroom, the town onto a world stage, the play of male and female power wrecks kingdoms.

Hugeness, intensity and purity of obsession are the last unchallenged heroic values. They flourish in worlds untouched by classical civilization, urban values or even female power. The brooding epic, full of fatalism and fury, is where every hero knows in advance that the largest triumph belongs to the grave. Going back to the wellspring of the Anglo-Saxon values James Page parodies, Gardner produced an extraordinarily brilliant novel, *Grendel,* a retelling of the Beowulf legend from the monster's point of view. This imaginative version of the legend in which Beowulf, the champion of civilization, kills Grendel, the embodiment of the terrors of night and nature, becomes a parable of the birth of modern man as a monster, as the scourge who dies.

In ringing language, Gardner turns the old story of man's necessary, if doomed, attempt to defeat the powers of destruction into a power struggle without ideological differences. Gardner's Grendel is evil as calculation, as a conscious choice not to believe in the value of love, in loyalty to any code of honor, or in any attempt to create order and self-esteem. But his Beowulf is a hissing angel of death, perhaps an agent of the malicious powers Grendel serves. He only arouses Grendel's desire to return to those gods of malignancy. Beowulf and Grendel are not adversaries, but conspirators. Grendel, wounded by Beowulf, chooses to jump from a cliff, to defeat death by joining it. He is heroic because of the sheer intensity of his blackness, because he allies himself with death, the all-powerful aggressor.

What happens when the primitive hero enters the age of the Christian knight? Heroism there depended on a chivalric code demanding that the great-hearted man be capable of worshiping woman as the helpless creature his might must protect. Gardner parodies this code in his modern, pastoral novels. Love as service to a lady, power as a tool to protect the weak, charity and free-flowing generosity are the knightly values ironically etched by Gardner in the devotion of the Sunlight Man to the sweet, gentle woman he marries only to find she is crazy, violent and convinced of her own omnipotence. As he labors to save her from madness, she sets fire to his office, burns his face, and destroys his career. Henry Soames in the beautiful novel *Nickel Mountain* is a big-hearted man who marries a young, pregnant girl, cares for her and loves her child as his own. But this modern Joseph has heart trouble and eats himself to death as he is displaced from his business by his wife. Coming of age, she turns out to be an ambitious businesswoman. The knight of charity finds himself done in by his mythology of the helpless princess. In defending the Equal Rights Amendment, does Sally display the failure of femininity that plunged the Christian knight to his grave?

Sally and James are past sex; for them, male and female are principles in a power play. Sally is Medeia as a media freak, the silly old woman whose best ideas come from the evening news. She is Gardner's example of the American woman as force and fury, the powerful, controlling principle hiding in liberal tolerance and instant sympathy for the underdog. Through her Gardner trivializes the values of the America she represents and elevates misogyny into an attack on assaultive femininity.

Gardner's men hate the new America and confuse it with

women. Like the smuggler Captain Fist in Sally's thriller, they have only two choices: "to turn on [women] and everything that reminded him of them with rage and scorn, or accept them, be swallowed up like the rest of us in effeminate softness and confusion—or give into a world so feminized that revolutionaries with slogans of death and home-made atomic bombs are softly analyzed, generously understood. Imagine a whole planet of big-boobed girl congressmen." James Page, raising his gun to kill a bear, thinks he hears his wife's pleading voice and shoots above the bear, wrecking a lifetime of shooting to hit the mark. Unlocking his heart, the female voice wrecks his "manliness" and makes him vulnerable.

Gardner hates softies. He does not settle on inward, inactive men like the heroes that fill recent fiction, but chooses action instead of passivity. Using one obsession to attack another, his novels move with a pacing and complexity that are remarkable in current fiction. But as these angry arguments spark each other, Gardner stands back. In not taking sides, he keeps his novels going and avoids the paralysis of action that marks many modern novels. But he arrives instead at a kind of moral paralysis.

Gardner is too much the product of the times he condemns. He is cynical about the aims of heroes from Beowulf to James Page, almost the first and last Anglo-Saxon ideals. Missing in his novels is a sense of masculine purposiveness, of the hero as the man who can put his power in the service of a worthwhile cause. Gardner's heroes are willful, self-absorbed narcissists who see determination and merit as the same, and whose heroism involves only the willingness to follow their obsessions wherever they might lead. By that logic Evel Knievel is no different than Neil Armstrong, and G. Gordon Liddy is every bit as grand as Winston Churchill. In not making distinctions between the value of one obsession and that of another, Gardner stalemates his novels.

Gardner's irresolution takes the form of an irony so pervasive it seems to stem from that well of American bitterness that made Herman Melville and Mark Twain, creators of distinctive American heroes, finally black about America's possibilities. His renegades from the modern world are figures out of that acid cartoon, Grant Woods's *American Gothic*. Gardner presses his ambivalences into *October Light* and the thriller it contains, forcing his chauvinism and his nihilism against each other like monuments to two American civilizations.

The heroes in novels of aggression by Mailer, Stone, and Gardner are men whose principles are formed not by political commitment but by political disaffection. This kind of radical hates other people's power over him, but may not necessarily indict America. He inherits a political climate in which faith in a communist utopia has already been destroyed by Stalinism and in which belief in any system is foreign. He sees his combat zone not as capitalism but as jungle warfare, as the war between the sexes or the junkie street fight in which the rip-him-off-before-he-rips-you-off law prevails. His surviving obsession is power. This hero is the system's product. He is not against the establishment so much as for himself.

Politics and narcissism are the same for the hero whose main code is "me first" and whose heroism lies in the intensity of his self-expression. If he is a right-winger like James Page, he may burn down his son's house in response to his son's failure; if he is a racist, he may burn crosses or tar blacks like Stone's crazies. If he is Norman Mailer, he may praise murder. The enemy the real world offers is less important than the hero's ever-permissible self-expression. The battle of Anopopei in *The Naked and the Dead* explodes even the illusion of American heroism. In the Vietnam War in *Dog Soldiers*, the only pure thing is heroin. Wars offer the opportunity for each man to confront his own inhibition, vulnerability and dependency. It is these he must destroy.

The confrontations in these novels highlight the changes in classical heroic values. In the *Hildebrandslied*, an extraordinary Germanic fragment, a father who has sworn loyalty to his chief finds on the battlefield that he must fight his own estranged son who has sworn loyalty to an enemy chief. Each must violate either the tie of blood or the tie of responsibility to survive. What happens is less important than the sharp sense of pain over violating either bond, since both are sources of meaning and pride. The new hero does not dread such a violation. He dreads the bond. The career marine turned pusher in Vietnam, his woman who abandons her child to go with him, are prototypical of the changes in attitudes toward political and human interconnection that have made every anchor seem like a trap. Fighting for a cause seems in these novels a form of outmoded social optimism.

The classical man of action acted to achieve a goal. The new one acts to perpetuate his disengagement, to feel like God. Heroic masculinity was once defined as authority, completion, the cessation of anxiety. Aeneas was among the first heroes whose options were painfully circumscribed by political necessity. His heroism lay in his ability to accept the limits on his will and his power and to fulfill his mission, the establishment of the Roman state. The evil he does to women is offset by his striving for a particular purpose, and secondarily by the fact that his abandonments bring him no pleasure. He is not an adventurer although his life is a series of "adventures." In Book III he remarks to Helenus at Epirus,

> Live in happiness, you are lucky to have fulfilled your destiny!
> We are thrown from one danger to another.
> You are not driven to plough the prairies of the sea, or to
> Chase the forever receding shores of Italy.

The chase that was tragic to Aeneas is the goal for the modern hero who wants to keep moving forever. The prize is simply the pursuit for the man who hates the catch, who idealizes the moment of combat (Mailer), the ability to be a detached manipulator of other people's anger, or the life of emotional invulnerability (Stone, Gardner).

The novel of aggression documents the personalization of political ideas, the degree to which the "system," the nation and the mistress can be used interchangeably as political symbols. Political labels (radical, conservative) apply more to personal behavior than to political ideas. Power in the novel is desirable more as an adornment, a form of gratification, than as a political fact, perhaps because the scale of aspiration in America is so great it operates to inflame self-hatred with whatever is actually achieved. This is a culture which produces a Nixon, a poor boy who could become president, but would then decide that was not enough. It also produces Rheinhardts, men who are objectively helpless yet achieve the illusion of omnipotence through refusing to play the game, through rejecting the ambition that engrosses the boss politician. They are opposite sides of the will to power. What is distinctive about this period is that the differences between real power and the illusion of omnipotence are continually blurred. The erosion of any clear sense of such distinctions operates in the novel to produce a false equation between what feels real and what is. Sensation becomes the standard by which truth is judged.

The revolutionary style, the sense of oneself as radical, has diffused so easily through American life because it is tied not to ideology but to the code of appetites by which the dominant culture operates. The revolution that has lasted is in behavior, in one's personal definition of the permissible. The family replaced politics as a battleground; the differences between politics and sex are, for novelists so unlike each other as Gardner and Stone, virtually eradicated. The primacy of women as the novelist's target (Mailer, Gardner) or as vehicles for expressing vulnerability (Stone) has made sex the measure of radical feeling. As Paul Theroux's hero Hood says in *The Family Arsenal* of his IRA terrorist lover, "Comrades became lovers, lovers conspirators, and promises whispered . . . in bed . . . evidence of political involvement." Love me, love my cause! for the cause is certainly me. The revolution is not between England and the IRA but between you and whoever or whatever is blocking you. It extends to all social strata. Theroux's cast of characters are aristocrats, actresses, urchins, diplomats, who rebel by taking up with their opposites. The English Lady is a bisexual terrorist sympathizer who loves delinquent girl orphans. The actress-activist wants to be Peter Pan. Kids revolt by blowing up statues of authoritative Englishmen like Lord Nelson. Theroux's revolutionaries live like a family in an ordinary house in a working-class residential community. But the point is not that terrorists are just family people so much as that the family is a hotbed of radical resentment. The upstairs room filled with guns is the skeleton in any family's closet, the bomb that will explode if the door of personal anger is opened.

The latest revolution is against anger, against the family bond of knotty emotion. If sex and violence were revolutionary for Mailer's hipsters, or for Gardner's Sunlight Man, the popularity of both have deprived them of radical meaning. In Stone's novels and in Theroux's *Family Arsenal,* the sexual revolution is a war of attrition too wearying to fight. Hood rolls opium pills and curls up next to his woman to dream of delights he will never get by touching her. The lovers in Stone's *Dog Soldiers* have sex only once and decide to shoot heroin instead. The slogan used to be, Make Love Not War. But the hero now increasingly wants to keep himself aloof from either, to make his assertiveness lead to the statement "No help!"

Novels of aggression obviously do not tell the whole truth about American radicals. The period in which these novels are set is one which saw the passage of extensive civil-rights legislation, the inception of the Peace Corps, and the beginnings of the peace movement that would become a crucial factor in ending the war in Southeast Asia. These significant accomplishments were not achieved by people who were as damaged as a Rheinhardt or as cynical as a dope pusher. The women's movement, which attempts to confront the sexual issues raised in these novels, is not even a possibility or a significant subject in them.

These novels do, however, tell truths. They document an atmosphere of self-interest, futility, cynicism and bitterness that fuel the criminal, the political boss and the fanatic to take a shot at wealth, fame, the illusion of godhood. They document the quality of appetite and narcissistic striving that drive people toward the gun or the syringe. Both the corporate president and the bank robber who pulls his job in time for the network news know how easy it is to feel like nothing in America. The erosion of the very obvious and very real distinctions between the businessman and thief, the bed and the economic structure, between the government and the local pusher, between actual power and illusory omnipotence, makes it possible for novelists to deal in an immediate way with the grandiosity and humiliation that are opposite sides of American life. In the novel all experience can be reduced to the ego trip, the fantasy of power shared by buyer and seller, revolutionary and boss and charmingly posed as a question by Theroux's pimp hero, Saint Jack: "Is there *anything* you want?"

VII

ON AND OFF THE TREADMILL

I am in the deepest sense an unfortunate individual who has from the earliest age been nailed fast to one suffering or another, to the very verge of insanity, which may have its deeper ground in a disproportion between my soul and my body.

Soren Kierkegaard,
The Last Years: Journals 1853–1855

To perceive life as a power struggle you cannot win, to see it reduced to cruelty or suffering, are familiar experiences in current fiction. What is striking is not that they are there, but that they are handled so successfully by novelists for whom victimization and violence are pervasive, binding conditions. Such writers are at home in disproportion, in the space between fiction and fact, between the dictates of social class, disease, and sex and free aspiration. Their holistic fiction seems an act of reconciliation between the world they have and the world they want.

Flannery O'Connor is a superb writer whose refusal to feel cornered brought her stories a force and authority that only increased as she fought a losing battle. This writer speaks to the condition of the child, the woman, the sick man whose weakness only marshals dreams of strength. Flannery O'Connor had a gift for silence. If women for centuries before her had concealed passion or fury in an outward calm, Flannery O'Connor turned that talent into a way of life. Her stillness was her refuge and her salvation. She grew up quietly in Milledgeville, Georgia, a patrician town that has not forgotten it was the state's last Confederate capital, a town that knew her as the daughter of Regina Cline O'Connor, whose family has, for more than a hundred years, been one of the most distinguished in the county. She politely remained an obedient daughter, one who kept her good humor when at twenty-five she was stricken with lupus, the wasting disease that confined her to her mother's farm for nearly fourteen unbroken years until it killed her. She agreeably became, for many readers, the Catholic who died cheerfully in her church. And in silence she wrote those quiet stories where violence so unexpectedly erupts, exploding all the values of obedience, politeness and faith.

The tension between O'Connor as Catholic daughter and Southern gentlewoman and O'Connor as writer bristles out of that stillness, the stillness in which she became a living contradiction: a woman who lived out a fiction and wrote down her life. Or at least her inmost life. O'Connor's silence thrives in the South, where women are taught from childhood to bury their passion or rage, to conceal inner turmoil behind a facade of feminine mildness. And in the dogwood groves and white-columned homes of Milledgeville, you can still find the result: the impenetrably sweet, unbelievably nice Southern belle.

O'Connor filled her short stories with such ladies in various stages of decline. One of them is the grandmother in "A Good Man Is Hard to Find," who yearns after the six-columned plantation she knew as a girl. She so wants to return to the house and the Old South it suggests that she tricks her son and his family into turning off the highway to Florida back into the Georgia hills. There they encounter the Misfit, a murderer who is also a gentleman. The Misfit apologizes for not being better dressed before a lady and, while his assistant murders her son and grandchildren in a nearby glen, he gets into a polite conversation with the grandmother. " 'You don't look a bit like you have common blood. I know you must come from nice people,' " she remarks as she hears the gunshots. " 'Yes ma'm,' " the Misfit agreeably replies, " 'God never made a finer woman than my mother and my daddy's heart was pure gold.' " But the Misfit is uncertain about what God did; he can't decide whether Christ really did raise the dead. Seeing his discomfort the grandmother says, " 'Why, you're one of my babies. You're one of my own children!' " Then,

"She reached out and touched him on the shoulder. The Misfit sprang back as if a snake had bitten him and shot her three times through the chest. Then he put his gun down on the ground and took off his glasses and began to clean them."

Was Flannery O'Connor Misfit or Lady? Or, rather, how did she manage to be both? I thought I could decide by going to Milledgeville and talking to people who knew her. But for the most patrician ladies in town, Flannery O'Connor is about as human as a Hepplewhite chest. She is displayed to advantage and said to contain the most neatly folded Southern virtues; she revered the traditions of her family, had beautiful manners and an equanimity so stable that when she was sixteen and her father died of lupus she could comfort her mother by reminding her that he was better off than they were.

On the farm where she wrote, raised peacocks and rocked on her porch until she died at thirty-nine, you can hear the geese bleat from time to time, or the sharp yowl of the peacocks. While I was sitting on the porch, listening to her mother's accounts of how happy her daughter was entertaining the admiring people who sought her out to talk, the silence only seemed to grow. Mrs. O'Connor prefers not to talk about O'Connor's work or her illness, nor will she permit anyone to see the specialist who treated her daughter. She seems to feel that talking of O'Connor's illness will "prolong it" and prefers to remember the serious, wry girl who has become famous. Nor did I feel Mrs. O'Connor's reticence came from an understandable desire to protect her daughter's privacy from a prying Northerner. It seemed to mark how content she was with her daughter's compliant exterior, how much she needed the doll her daughter had created for her, and how skillfully O'Connor had hidden her complexity.

But the actions of Flannery O'Connor have an eloquence all their own. She left Milledgeville to become a writer "on her own," as Robert Fitzgerald, her friend and literary executor, says in his introduction to the last collection of her stories, *Everything That Rises Must Converge*. She went to the University of Iowa for a master's degree in writing and then came to New

York, not a surprising move for a woman of twenty-three who wants to be anything "on her own." What is surprising is that she so quickly left New York to live on an isolated Connecticut farm as the boarder of the Fitzgerald family. The "shy, glum girl" Fitzgerald describes seems to have been frightened by freedom, seems to have needed the peculiar isolation and comfort that comes from friendships maintained through letters or kept with married friends. This only child of a self-consciously aristocratic family, praised by her mother for not seeking out friends, seems never to have been deeply close to anyone.

O'Connor's life on the Connecticut farm, devoted to days of writing and evenings of conversation with Sally and Robert Fitzgerald, was the life she chose for herself when still "on her own." It was not so very different from the life her illness would impose on her. In fact, I do not think O'Connor's disease radically changed her life. Its very horror was that it prevented her life from changing at all. The loneliness it dictated for her was all too familiar to the "shy, glum girl" whose feelings had always been so under control, who seemed so essentially alone everywhere. Her illness seems only to have cemented an isolation that had always existed, a feeling of being "other" that she could at times accept with wry good humor. I am thinking in particular of a cartoon she did as an undergraduate at the Georgia Women's College. It shows a girl who looks like O'Connor wearing huge eyeglasses and sitting alone at a dance while couples swirl all around her. She has a desperately cheerful smile. "Oh well," the caption reads, "I can always be a Ph.D."

If O'Connor joked about what bothered her as an undergraduate, she would later mask a greater pain in frenetic good cheer, treating her weakness and disfigurement with humor and turning herself into a cartoon character. "I am doing fairly well these days, though I am practically baldheaded on top and have a watermelon face," she wrote to the Fitzgeralds at twenty-eight. At thirty-eight, some twelve weeks before her death, she wrote to Richard Stern "a letter scarcely different from any I'd received from her in the five years we'd known each other except that the signature was pencilled and shaky." The letter reads:

Milledgeville
(April 14)
1964

Dear Richard,

I'm cheered my Chicago agent is keeping up with his duty to keep you informed on my state of being. It ain't much but I'm able to take nourishment and participate in a few . . . rallies. You're that much better off than me, scrapping Tuesday what you wrote Monday. All I've written this year have been a few letters. I have a little contribution to human understanding in the Spring Sewanee but I wrote that last year. You might read something called GOGOL'S WIFE if you haven't already—by one of those Eyetalians, I forget which. As for me I don't read anything but the newspaper and the Bible. Everybody else did that it would be a better world.

Our springs done come and gone. It is summer here. My muscovy duck is setting under the back steps. I have two new swans who sit on the grass and converse with each other in low tones while the peacocks scream and holler. You just ought to leave that place you teach at and come teach in one of our excellent military colleges or female academies where you could get something good to eat. One of these days you will see the light and I'll be the first to shake your hand. . . .

Cheers and thanks for thinking of me. I think of you often in that cold place among them interleckshuls.

Flannery *

O'Connor's cheeriness in the face of her own dissolution is, as most people say, admirable. But beyond her cheeriness is the tragedy bound up with that very courage. For O'Connor's cheeriness also expresses her older, more essential malaise—that deep isolation that seems to have made her happiest when alone in an insulated and protected world, in a world

* Richard Stern, "Flannery O'Connor: A Remembrance and some Letters," *Shenandoah*, Vol. XVI, No. 2 (Winter 1965), pp. 5–6.

where the most constant and enduring attachment is to swans, ducks and peacocks. If her letter is an affirmation of life at all, it is an affirmation of the life of a very lonely woman. Her illness seems to have made it possible, perhaps essential, that she play this role: the country wit who likes to poke fun at complexity and Northern "interleckshuls." This tough, rather boyish, cheery idiom seems to have been one of her salvations. It seems to me to be a distinctively Southern salvation.

What Flannery O'Connor took from her background was the ability to disappear into her behavior—to become the role she played to such an extent that she could detach herself from her own pain. In all her cheerful patter you miss entirely the suffering that must have been its ultimate source. The country diction is oddly mute about the anguish of a woman feeling the slow violence of disease. It is in fact so inexpressive of anything humanly true that its silence becomes eloquent.

Whatever happened to her or whatever she felt, Flannery O'Connor seems to have followed the code quite rigidly. Not the code of the Catholic Church but the more rigorous code of genteel Southern womanhood, the code that *forbids* confession. This is the code she would have heard from childhood in those wonderful sayings: "Pretty is as pretty does," or more threateningly, "Don't fuss!" Or, as Regina Cline O'Connor put it, "I was brought up to be nice to everyone and not to tell anyone my business." Being "nice to everyone" is what produces the immensely attractive surface of Southern life, that faultless politeness and sweetness that you can still encounter often enough in Milledgeville. Yet it is a politeness that engulfs every other emotion, that permits no contact on any but the most superficial level. And it is not merely a graciousness that is carted out for strangers. It seems to exist between mothers and daughters, husbands and wives. It seems everywhere to substitute "doing pretty" for genuine warmth. It censures all of those excesses of feeling that would constitute "fussing" and prohibits being close enough to anyone to tell him your "business."

But what if one's "business," one's most essential feelings, are not the stuff pretty gestures are made of? What if, from girlhood, you have known you loathe the Southern belle you are supposed to become? What if you have felt "other" and "different" in a milieu that is horribly embarrassed by anything unconventional? And what if your "business" later on is dying slowly, being filled with impotent rage at your own weakness? And what if, through it all, no one will even tolerate you "fussing" about it?

Like her mother, like the tight-lipped kids in her books, Flannery O'Connor, I suspect, told her "business" to no one. She rather retreated from it herself into the role of a witty Georgia girl. It is not surprising that, after a particularly horrible bout of her disease, she would paint a portrait of herself next to a peacock looking like a suffering, proud adolescent in a wide straw hat. Like Mrs. Hopewell in "Good Country People," who thinks of her daughter as a child because it is too painful to think of her as a thirty-two-year-old Ph.D. who has had a weak heart and a wooden leg for twenty years, O'Connor seems to have found a way out of her unpretty situation by denying it.

O'Connor filled her stories with women who are always kids, or men who are bound to their mothers in infantile need. She never wrote of sex except as some unknowable perversion. When Joy-Hulga tries to corrupt the Bible salesman by seducing him, she finds he has no interest in anything but stealing her wooden leg and learns that he acquired a glass eye through another woman's expectations. In "The Comforts of Home," Thomas is repelled by the advent of Star, a nymphomaniac his mother pities and takes in, because she interferes with his solitary bliss with his mother. O'Connor's sons and daughters are their own Mrs. Hopewells, regarding themselves as though they have no sex at all and living in denial of need.

The fiction O'Connor lived had its roots in that Southern need to do pretty regardless of what you feel, and in her own remarkable ability to divorce behavior from feeling. Much as she scorned the Mrs. Hopewells, Mrs. Mays, and Mrs. Turpins she wrote about, she was, in many ways, like them. She denied in her own life what they denied in the lives of their daughters and sons. She seems to have found an alternative to the Southern belle in a variety of roles, ranging from the witty Georgian to the uncompromising Catholic. She seems to have gone through the motions of conventional behavior without becoming deeply involved in the conventional world around her and without expecting any deep human contact. She was compliant, affirmative, happy with farm life, an excellent daughter, a Catholic unresentful of death; but she seems to have been oddly out of touch with those more essential feelings that explode in her work. And it is through her very ability to detach herself from those feelings that she came closest to being what she had never admired: a Southern lady.

Milledgeville is publicly proud of Regina Cline O'Connor's daughter. In the public library you will find a memorial display of her work accompanied by a vase of peacock feathers and a curiously out-of-place painting O'Connor did of an enraged rooster. Freshmen at her alma mater, the Georgia Women's College, which has, since O'Conner's time, changed its name and admitted men, are required to read her work. Yet despite the outward display and approval, you will find that almost no one withdraws her books from the library and few people buy them. When they do, it is usually as gifts for people up north or out of town. They ship them off as genuine Milledgeville products, often because Atlanta critics or Northern "interleckshuls" have said they are admirable. Many who have read some of O'Connor's work do not read more. They apologize for not being more "literary" and explain that they cannot rid themselves of the feeling that there is something very "peculiar" about it, something very "different."

For some who knew her, Flannery O'Connor *was* "different." There is in the memory of one Milledgeville matron the image of O'Connor at nineteen or twenty who, when invited to a wedding shower for an old family friend, remained standing, her back pressed against the wall, scowling at the group of women who had sat down to lunch. Neither the devil nor her mother could make her say yes to this fiercely gracious female society, but Flannery O'Connor could not say no even in a whisper. She could not refuse the invitation, but she would not accept it either. She did not exactly "fuss" but neither did she "do pretty." Perhaps the essence of Flannery O'Connor is precisely there, in that mute scowl. Maybe that is what made her "different."

Perhaps there is something of this rage even in O'Connor's love for peacocks. Did she admire the ease with which they gobbled up all the flowers in sight, destroying her mother's flower beds and turning the lawn white with droppings? Were those majestic birds that broke all the rules what Flannery O'Connor wanted to be? The curse on the bird is its yowl—the ugly voice that makes it most beautiful when silent. But to Flannery O'Connor that voice sounded like "cheers for an invisible parade." Was that parade the procession of Misfits, prophets, and lonely and murderous children who unleash their violence so freely in the fiction of Flannery O'Connor?

Flannery O'Connor never yowled in public. She never said what her mute scowl expressed. But she rendered it in pictures as powerful as the tableau of the grandmother and the Misfit, bound in silence to each other through a ritual of politeness. The Misfit can find no words to speak his rage at his would-be mother. Fury explodes from his gun in three eloquent shots. And one of the revelations in "Revelation" is Mary Grace's peculiar wrath. As her mother and Mrs. Turpin criticize her, Mary Grace, a fat, pimply girl who never smiles (she got ugly up north at Wellesley), accepts her mother's remarks politely but grows enraged at Mrs. Turpin. She gets so angry she throws a book at her. But after she hurls it, her very fury makes her crumple to the floor. Her mother bends over her, a doctor sedates her while she clings irresistibly to her mother's hand. O'Connor perfectly welds her need for her mother and her hatred of her, meshes them into one experience of the destruction, the humiliation of needing someone who refuses to accept you at all. And Mary Grace is too vulnerable, too crippled even to attack her own mother. She attacks her mother's "double," Mrs. Turpin, while leaving her own mother alone, much in the way the Misfit claims there was "no finer" woman than his mother, but goes on to murder a woman who suggests to him all the forces of tradition and family and claims he is "one of her babies." And even the Misfit is left wiping tears from his glasses.

O'Connor's murderous, vulnerable children are always "ladies" and "gentlemen." They always say the right thing or nothing at all; they behave properly to their parents. But they are always furious at the parents who have made them so polite, or who try to destroy their pride in being misfits. Some have a secret inner world where they never obey. And in quiet acts of violence, others unleash their fury.

O'Connor wrote about what she knew best: what it means to be a living contradiction. For her it meant an eternal cheeriness and suffering; graciousness and fear of human contact, acquiescence and enduring fury. Whether through some great effort of the will or through some more mysterious and unconscious force, she created from that strife a powerful art, an art that was both a release from and a vindication for her life. If she set out to make morals, to praise the old values, she ended by engulfing all of them in an icy violence. If she began by mocking or damning her murderous heroes, she ended by exalting them. Flannery O'Connor became more and more the pure poet of the Misfit, the damaged daughter, the psychic cripple—of all of those who are martyred by silent fury and redeemed through violence.

Joyce Carol Oates can be as savage as she is rhapsodic. In *them* she brought to the Detroit riots of 1967 the special inner power and rightness O'Connor could bring to the disputes between mothers and daughters. In Oates's novel, the riot is not a simple power play between the rich and the poor, but the product of differences between generations and of the diffusion of rebelliousness throughout American life.

The parents of the young people who make the riot in *them* live in a world of sexual violence in which men have been undermined by Depression poverty and by the hard and hardening sense of failure of the working poor. The novel opens with Loretta in love with her reflection, primping for a Saturday night date that will result in the murder of her first lover by her brother, her terror over the murder and her gratitude toward Howard, a policeman who helps her get rid of the body in return for a sexual shot at her. Men to Loretta are "silent and angry, hungry but impatient with food, pushing it around on a plate, stuck with a terrible burden of flesh and needing someone like Loretta to ease it." Women, on the other hand, talk constantly to each other, have a comradeship based on babies, on "sifting, judging, preparing." Neither this kind of male anger nor the sexual dependency of a lusty woman is reproduced in the children Howard and Loretta have. Loretta's sexuality, her ability to give in to life, is completely beyond the uptight Maureen who, after years as a dutiful Catholic schoolgirl, begins primly picking up men and charging them for sex. She feels nothing but interest in the money they give her. When her father beats her for this, she retreats into silence and food, consuming quantities of sweets and speaking to no one for more than a year.

Jules is brighter and more emotionally sensitive than his father. He is set on the right path. He falls deeply in love with a rich, married woman who seeks him out, rents an apartment where they can meet, and wants him to give her a sense of being alive. She shoots him when after an afternoon with him she has not been able to have an orgasm. He gives up trying to be Horatio Alger or even Gatsby, and falls in with university radicals who are fascinated by his lower-class look and show him how his bitterness can be socially useful. In the climate of sixties tolerance, he even gets radio time to deliver his message, the lesson experience has given him: "Fire burns and does its work."

Maureen, emerging from her fat and depression, finds her own route to violence in entangling a married college instructor with three children in an affair with her. She feels nothing for him but knows that if she can persuade him to leave his family, it will be a sign of his love for her and he will never have the courage or money to make another move. Like her mother, whose one proof of her former power is that a man was murdered for her, Maureen proves herself by ruining a man's life. She succeeds. At the end of the novel she has no feeling for her new husband but rushes her activist brother out because she is eager to seem a "normal" wife and unwilling to have her husband meet him and discover the destructiveness in her family.

Loretta and Howard were content to be victims of their class and of each other. Maureen and Jules, instead of being beaten into acquiescence by the hardness of their lives, are fueled for revolt. That each of them acts destructively and self-destructively is less important than Oates's sense that what divides generations is the willingness to return the blow. What divides the sexes is how they hurt each other—what defines the times in *them* is the fact that authoritative institutions (the university, the media) are so permissive that they make it possible for the violent man to become a social force.

The subtlety of *them* comes through even its sentimentality, in Oates's ability to be so many voices at once, to catch the sound of working-class speech, to get into the head of the woman in a housedress talking with her neighbor in front of a mailbox. Her ability to catch domestic details and to make them monuments to domestic stability gives this novel its remarkable power. The fact that the significant characters are poor whites may distort the impact of the Detroit riots for blacks, but nevertheless permits the novel to remain within bounds familiar to a writer who came from a poor rural community.

Oates is no radical. But she is the young American novelist closest to the tradition of social realism. Her subject is not the goodness of the suffering poor but the extent to which being poor in America creates a divided soul. In *A Garden of Earthly Delights* and *Expensive People* she makes an equation between poverty and the violence of women who can pass as normal middle-class and upper-middle-class wives but who, by force of earlier economic deprivations, are witches incapable of any real feeling and invariably damaging to their men.

The other side of the witch is the masochist who, because of the wrongs done to her, languishes through life waiting for her pain to end, her condition to change. *Do With Me What You Will* and *Wonderland* explore this theme. But in *Childwold* Oates reaches for the origin of the sense of unlovability that afflicts her heroines. This novel of a poor rural family—a blowsy mother whose many children have many fathers, her fourteen-year-old daughter Laney, and Kasch, an anguished intellectual who loves them both—extends Oates's fascination with poverty as the dictator of personality. Laney has a horrible sense of freakishness, of being an outcast, that comes only partly from her sense of distance from her mother's sensuality, her mother's ability, like Loretta's, simply to give in to an environment. Laney's most acute pain is the product of Oates's acute perception of poverty as a form of exclusion and self-exclusion from the world.

What links Oates to the social realists of former times is her ability to impress into the personal anxiety of an adolescent the largest social awareness. Laney, taken to a gallery by wealthy Kasch, thinks: "Everything in this place has meaning, people have come here to experience the meaning, they know it is here, it has been deliberately and lovingly created and so they have come here, have journeyed here, knowing they will not be disappointed. . . ." In a store where Kasch wants to buy her an expensive coat, she looks at the price tag: "You are shocked and angry and want to tell everyone in the store to go to hell, . . . how dare they, you feel sick, you feel as if people are watching you and about to burst into laughter. . . . it's hopeless, forget it, forget everything. . . . ." This is poverty as personality, as an insecurity big enough to drain every day of possibility.

An impoverished childhood seems better to Laney than her probable future as a woman. Watching her mother drift from man to man or being beaten by her lovers, Laney is persuaded that sex reduces women to sluts and men to brutes. Kasch, who seems so tender in conversation, sees sex with Laney as violence: "That night I grappled with her and overcame her and snapped her fragile bones in my heated love. . . . I tore at her mouth with my own. I tore at her tiny breasts, her thighs." Unable to renew himself with Laney, Kasch is easily seduced by her more experienced mother. But reviving his appetites only inflames his rage. He murders one of his rivals so brutally that he goes mad. Oates's women come of age and get sexually mugged; her best men are turned by sheer excitement into every-ready fists.

Childhood is the fragile barrier against the future as a have-not or a killer. The novel's tight Oedipal triangle opens into a triple alliance against age and aggression as each person tries to turn the biological clock back towards innocence. Laney's mother wants to bear children to narrow the world to a child's room. Laney starves herself to stop her menstrual cycle and prolong her childhood. Kasch in his insanity is harmless, non-sexual, helpless. Outside of sex, Laney and Kasch may connect. Laney calls Kasch back from his insanity: "Bearded, gaunt, sickly-pale hair colorless as dead hair, eyes in shadow . . . is it Kasch, here, is it Laney, here, shading her eyes, waiting for a sign?" Drawn by his vulnerability in the same way that Kasch was drawn by her poverty, Laney may cling to Kasch and he to her as children cling together against the dark.

Oates bares characters who are driven by blind impulse through the pitifully few parts they can play. Oates rhapsodizes the incomprehensibility of their lives. She virtually pleads for their blindness as a way of not seeing how little real mystery there is in lives that seem predestined to be unhappy. Fate could have written this primal drama of mother, daughter and rich man. And the novel's major flaw is exactly an almost superhuman, torrential flow of words that washes out the individual voice and often makes it difficult to tell who is saying what. Oates's characters are frequently accretions of sex and class, victims of a violence they recollect with nostalgia.

Protective, sacrificial love as a weapon against greater despair has gone out of fashion, particularly in the novel. In *them* Maureen, who typifies the new generation, strikes back at men by ensnaring one she does not love. In the explicitly feminist novels of Marge Piercy like *Small Changes,* the heroine can walk out of a marriage or even become a lesbian. Oates most often turns back to the heartland of American fiction. This insistent writer can call us back, through her dense psychological and social detail, to rural America in a spiritual and economic depression. The characters of *them*, of *Childwold*, and of *Garden of Earthly Delights* inspire the terror and recognition once aroused by those monumental lonelies of William Faulkner, John Steinbeck and Carson McCullers.

Passing over the current sense of sex as a matter of problem-solving, Oates turns back the clock toward a Christian tradition in which life is decipherable only through the code of compassion. Her answer to fire and brimstone, to the destructiveness that "does its work," is to reach beyond it to its source in human suffering and to praise redemptive love. Her novels are ultimately holistic, remaining fixed on lives which alternate between cruelty and masochism, between giving and receiving violence. Yet Oates as a novelist has mastered the code of compassion for both her perpetrators and victims and by force of her mastery she has penetrated her view of life as a life sentence.

Joyce Carol Oates's people hover between the impulse to destroy and a euphoric, tragic acceptance of their own pain at feeling destroyed by others. The force of her characters who do strike back is almost always in proportion to the wrongs done to them. Like O'Connor's Misfit who wishes to make "what all I done wrong equal what all I gone through in punishment," they are looking for a balance point. Neither O'Connor nor Oates equates the suffering of their women characters with being women; rather they use biological facts of sex, illness or age to express a larger vulnerability to both rage and despair. O'Connor moved unhesitatingly as a writer toward violence. Oates vacillates. At the core of their very different novels is a common struggle with a vision of life as a power struggle—between the weak and the strong, the affluent and the have-not.

One of the most influential women novelists to perceive life as an adversary relation is not an American. Doris Lessing's early interest in revolutionary politics, her ability to wedge women's lives into the mold of larger political forces, and perhaps her very distance from American feminism have made her a special, almost sibylline figure to her American audience. The personalization of politics and even religion was carried into another realm by the woman who was virtually the first and the most complex novelist to describe the problems of being alive in terms of the problems of being a woman. The heavy reflectiveness of Lessing's writing both separates her from most American novelists and seems only to draw her readers further through an oeuvre so rich in intellectual energy that it continually holds out the promise of solving the heroine's troubles.

Doris Lessing's heroines are always intense, bright women for whom life is a succession of traps created by their mothers, lovers and finally by themselves, by personalities that have the habit of confinement. Her purest talent has been for the passions of jailed lives—the desperate loves, intense rationalism, intellectual energy that all go reeling in the most rebellious directions. Intelligent and tough, sensual and eager for life, Lessing's heroines have everything but the freedom to be successful in one area of life without paying for it in another. Her most intelligent women are mindless in love, so unable to set limits that they are easily controlled by their men, or by anyone else's expectations. They let themselves sink into the most abysmal misery, a wretchedness so total it destroys their ability to feel or react at all. They may bolt, dumping their husbands and even their children. But once free, they do it all over again.

Despite the intelligence of Lessing's heroines and the reflective style of her work, her characters never gain enough insight into themselves to get off the treadmill of submission, rage and guilt. They think with the logic of the psychically jailed, for whom even thought is a process as closed as a knot. They retreat from understanding into great abstractions. This is not sexual politics so much as politicized sex. As Martha Quest, heroine of the five-volume *Children of Violence* series, grows stupefied and apathetic in her pregnancy and motherhood, she feels more and more for the plight of the world's oppressed workers and joins the Communist Party. But she never really confronts her husband or the sources of her own paralysis. Women communists like Anna and Molly in *The Golden Notebook* want to liberate themselves more than the workers. They grow cynical about the possibilities of world revolution while they grow contemptuous of how conventional communist men are with them. Where are the free men for the free women, Anna bitterly asks Molly.

Doris Lessing's work is not about ideology but about ideologues, the personalities of people who need a climate of abstraction, for whom politics is the soul writ large. No one has so effectively welded political dreams with sexual fantasies, no one has more successfully made great social facts and private anguish converge in the single, tense charge of probed experience—the anguish of women doomed never to face their brokenness, never to say I want, I need, but to conceal themselves in dogma, in marriage, in some engulfing force. They run to politics for relief from personal turmoil but find, in their rationalism, other traps.

Martha Quest at fifteen is miserably convinced that her life is poisoned. In *Martha Quest* she begins looking for a reason why from the psychologists:

> There was the group which stated that her life was already determined when she still crouched sightless in the womb of Mrs. Quest. . . . She rocked in the waters of ancient seas, her ears lulled by the rhythm of the tides. But these tides, the pulsing blood of Mrs. Quest, sang no uncertain messages to Martha but songs of anger, or love, or fear of resentment, which sank into the passive brain of the infant, like a doom. Then there were those who said it was the birth itself which set Martha on a fated road. It was during the long night of terror, the night of the difficult birth, when the womb of Mrs. Quest convulsed and fought to expel its burden through the unwilling gates of bone (for Mrs. Quest was rather old to bear a first child). It was during that birth, from which Martha emerged shocked and weary, her face temporarily scarred purple from the forceps, that her character and therefore her life were determined for her.

Lessing has written sensitive accounts of what goes wrong between mothers and daughters. Martha feels that she is always liable to sink into a total passivity where she has no eyes, no mind, no will, because she is battered between her mother's love and her ridicule. When she is sixteen, about to take the university exams that will bring her the life she wants, her mother persuades her to come home instead, using a minor eye irritation as proof that Martha could not pass. Once she gets Martha home, her mother ridicules her for not taking the exams. Martha keeps asking:

> Why was she condemning herself to live on this farm, which more than anything in the world she wanted to leave? She felt as if some kind of spell had been put on her. Martha was turned in on herself, in a heavy, trancelike state. What was so frightening was this feeling of being drugged . . . she did not understand why she was acting against her will, her intellect, everything she believed. It was as if her body and brain were numbed.

Lessing's account of Martha's adolescence is a shattering description of a mother's power to narcotize, of Martha's absorption of May's will. "You must be tired, darling," May says to her daughter. "Don't overtire yourself, dear." Although Martha angrily snaps back, " 'I will not be tired,' she felt herself claimed by the nightmare and, in fact, at the word 'tired' she felt herself tired and had to shake herself."

The only security Martha has known is her mother's belittling, overpowering control. Self-hatred and self-abnegation are the price of Marth's survival in childhood; impotent, guilt-ridden rage is the toll she pays for a lifetime. She leaves home for the city and her own apartment. Her mother visits her and arranges all her clothes in her dresser while she is out. Martha returns and angrily throws her clothes on the floor. But hours later, her mother long gone, she puts them back in precisely the order in which her mother had arranged them. She has so internalized her mother's malice that she kills her own happiness. She marries a man her mother has chosen for her. She discounts how miserable and depressed she feels because the "experts" say such feelings are "normal." Martha's anger contains self-punishment: she hates her husband so much she thinks even his sperm do not have what it takes and, neglecting to use a contraceptive, she becomes pregnant.

"I am my mother," wrote Anne Sexton. Martha grudgingly becomes the mother she hates. She sees her daughter, Caroline, as a pesty misfit who will not conform to the feeding schedule the "experts" have established, and tries to control her without treating her as a person or acknowledging her needs. She hates herself, believing she is as rotten a mother as May. She tells Caroline, "Parents should be suitable objects of hate: if psychology doesn't mean that, it means nothing. Well, then, so it's right and proper you should hate my guts. . . ." But she listens passively when May looks at Caroline and remarks with satisfaction, "I suppose you've been starving her as I've starved you."

Martha walks out when Caroline is five. She tells her she is setting her free. But May, who has ingratiated herself with Martha's husband, is happy to take over. Martha goes on to make another bad marriage. In *The Four-Gated City*, Lessing's best novel, she is well into her forties, far from her mother and her daughter, but a living demonstration that what women get out of masochism is structure, order, security. Martha is raising her lover's child, helping with his work and caring for his insane wife besides. Her greatest intimacy is with this madwoman whose disintegration helps her express and control her own. Martha buries her problems by managing other people's craziness, by living an uptight life. But a visit from her mother destroys her self-control and what happiness she has. Her rage surfaces, directed against herself. She withdraws from her lover, from herself as rationalist, and, in a shattering process of self-exploration, she drives herself mad. In her insanity she plugs into a universe of hating voices, encounters what Lessing calls the "self-hater." This freakish image of herself, this noisy metamorphosing monster, is the figure of Martha meshing with May, becoming her raging, negating mother. The self-hater is the cannibal within, the self that drives you to misery. But this disintegration is, ironically, the way to peace.

For Doris Lessing the road seems to have been through the wildest disorder. Lessing herself chose to enter insanity's abyss. As she told me in an interview,

> It's very easy to send oneself round the bend for a couple of days. I did it once out of curiosity. I do not recommend that anyone should do it. I'm a fairly tough character and I've been in contact with a very large number of people who've been crazy, and I know quite a lot about it, and I knew exactly what I was doing. I sent myself round the bend by the simple expedient of not eating and not sleeping for a bit. I instantly encountered this figure I call the "self-hater."

Corrosive and negating, the self-hater is the internalized voice of a mother belittling her child and coercing it into "the ways she finds it least offensive." Lessing seems to relate this inner voice to "another dimension which is very close to the one that we're used to and that people under stress open doors to."

In *Briefing for a Descent into Hell,* Lessing describes that dimension through the mind of a mental patient who, in his madness, sees his apelike brothers copulating and feasting on bloody meat. Nauseated, he "purifies" himself and prays for salvation and escape in a "radiant crystal" from outer space. The images are bizarre, the novel free and peculiarly primitive. But though mad and male, though a wanderer through space and time, Lessing's hero is not unlike her Marthas and Annas and Mollys, not unlike those women for whom freedom does not exist on earth, but only in an otherworldly form—as a dream of utopian communism or a spaceship filled with alien intelligences, unknowable sexes. And, Lessing seems to ask, what matter questions of male and female when every earthling is loathsome? Does Lessing believe in radiant crystals? Martians among us? She accepts them as possibilities: "It wouldn't take much, would it, for somebody . . . for intelligent creatures to disguise themselves in such a way that they would not be recognized?" Lessing believes in outsiders. And Martha Quest becomes happiest as a stranger.

Martha emerges from her madness believing more and more in the powers of evening, the mysteries of darkness. The novel ends. But in an epilogue, Lessing describes an apocalyptic third world war which destroys civilization and leaves Martha on an island, far from her loved ones. And Martha refuses to be rescued. This global war is the image of Martha's fury, the analogue of the man's revulsion at humankind. That war is like some explosion of the inmost self, the free unleashing of all the feelings Martha has always turned against herself or muted into an outward contempt.

But madness, war, and images of the basic loathsomeness of earthlings erupted through Lessing's works, only to cool. Doris Lessing praises age and the redemptions of time, and celebrates the erosion of need. She has given up on the promise of people and politics and belittles the importance of having needed them at all. "When you get to be middle-aged, which I am, it is very common to look back and to think that a lot of the sound and fury one's been involved in was not that necessary. There is quite often a sense of enormous relief, of having emerged from a great welter of emotionalism." And having emerged, Lessing has embraced "Biology," an ideology of age.

Lessing's Biology erases everyone to anonymous physical craving:

> We're very biological animals. We always tend to think that if one is in a violent state of emotional need, that it is our unique emotional need or state, when in matter of fact it's probably just the emotions of a young woman whose body is demanding that she have children. It's hard for many people to take, but 90 percent of our marvelous, unique and wonderful thoughts are in fact expressions of whatever state or human stage we're in. . . . Anna and Molly are women who are conditioned to be one way who are trying to be another. I know a lot of girls who don't want to get married and who don't want to have children. And very vocal they are about it. Well they're trying to cheat on their biology. And I say it will be nice to see what happens. It will be interesting to see how they're thinking at thirty.

Lessing's Biology ignores the plenitude of human personality for one vision of passive failure. For in this economy we are all biology's beasts of burden, whipped on by the torments of "youth" and reproduction that are eased only with age and climacteric. And we owe even our successes to Biology, to some mystical menopause that ushers in the self-knowledge that the mind and will could never achieve. For time presumably makes motherhood, daughterhood, even womanhood irrelevant.

> An attractive young woman finds it very hard to separate what she really is from her appearance. Because you only begin to discover the difference between what you really are, your real self, and your appearance when you get a bit older, which is the most fascinating experience. It really is. It's one of the most valuable experiences I personally have ever had. A whole dimension of life suddenly slides away and you realize that what in fact you've been using to get attention has been what you look like, sex appeal, or something. It has nothing to do with you. It's a biological thing. It's totally and absolutely impersonal. It has nothing to do with you. It really is a most salutory and fascinating experience to go through, shedding it all.

Doris Lessing the woman seems concealed in generalizations about womankind, wrapped in the most protective ideas. The intellectual energy that marked Lessing's earlier works now seems to bolster what her heroines once sought to escape: the exhaustion, passivity, abjectness in the face of their own experience that they once saw as the cause of their anguish.

Is happiness accepting biological necessity and death or is age the safest abstraction, the pursuit of self through the destruction of self? *Summer Before the Dark* is a novel of the liberations, the blessings of middle age. In this novel Lessing's heroine is hit by time with the force of a new faith. And she ages into salvation. A sensual woman who is a wife, a mother, a soother, Kate Brown goes off to Spain with a young lover. But she falls into a fever that is more ecstasy than disease and awakes to find that all her old passions have been burned out. And full in the ruins of her pleasure, she finds herself whole, elated and free.

Kate Brown rises from her fever transfigured. No longer the yielding, soft woman who was easily engulfed by other people's demands, she has somehow burned away her desire to please, killed her need for other people. In a baggy dress, her hair undone, she walks through her old neighborhood glittering with contempt for the friends who cannot even recognize her: "She was delighted, she felt quite drunk with relief, that friendship, ties, 'knowing people' were so easily disproved." At the theater watching Turgenev's *A Month in the Country*, she rises to a radiant disgust. As Natalya Petrovna, "the mirror of every woman in the audience who has been the center of attention and now sees her power slip away from her," agonizes over her love for a young student, Kate, who has only days before gone off with a much younger man, gloriously thinks: "We ought to be throwing rotten fruit. . . ." She mutters: "Oh nonsense, nonsense . . . couldn't anyone *see* that what they were all watching was the behavior of maniacs? Really they all ought to be falling about, roaring with laughter instead of feeling intelligent sympathy at these ridiculous absurd meaningless problems."

Freedom requires, Lessing seems to say, this repudiation of desire. Looking back on her life, Kate sees marriage and love, sex and friendship, motherhood, work success and making money collapse into one ashen glimpse of life as an unbroken fraud. Like Martha, who senses at the end of *Four-Gated City* that her future is bound up with the gathering darkness of evening, Kate has no clear vision of herself apart, no way to shape her life for the future, no means of escaping the rage that still binds her to the memory of what she was. And like Martha who thrives only on her island, Kate can flourish only alone and apart on the private, narrow islands of sleep, in dreams of freedom.

With an absolute sympathy, Doris Lessing writes of Kate's dreams of a seal she finds damaged and adrift on a hill far from the sea. Through dream after dream Kate carries it in her arms, protecting it against every disaster. The dreams all vibrate with those inviolate relations between self and self. For the seal is Kate, mother and wife, a helpless, broken animal locked, sealed in an alien world, a literal fish out of water, a cold fish craving some clear northern sea. Far heavier than any albatross, this seal is Kate's self as pure burden. In her last dream, she sets the seal down safely in the sea, watching it disappear among other heedless beasts. In her dream, Kate is lightened, finally delivered of this perpetual, damaged child after a lifetime of labor. This seal is the last child of violence. Its abandonment puts an end to turmoil, youth, attachment, vulnerability. If in her waking life all Kate can do is angrily mutter, put people off with her looks, put them down with her contempt, in sleep and in isolation she feels cleansed of her need.

Kate takes age as the rationale for her pervasive repudiation and withdrawal. Lessing is less concerned with personal realization in the world than with the pursuit of peace and quiet. Where Martha found herself through madness, conceived of as a universal well of hate, Kate finds herself through an impersonal physical process of age that affects every living thing. From youth to menopause, Lessing's women always withdraw from themselves, become as their mothers were to them, dehumanizers. No stylist, no elegant writer, Lessing has nevertheless written the most powerful portraits of people who are driven to the anonymous abstractions of reason or madness or death as the only available relief from the narrow particular of themselves.

Doris Lessing is a supreme novelist, the first and still the best to deal with women like Martha and Kate. She has gone beyond fury at the mothers, the husbands, the neuroses that keep her women down. If she no longer writes of the rage Martha felt for her mother for dispensing "sleep and death . . . like a sweet and poisonous cloud of forgetfulness," she writes effectively of sleep as the dream of fullness, the paradigm of total escape, absolute freedom in absolute isolation. Kate discovers that weariness is a barricade. And she fulfills an old intuition. Twenty years ago, Doris Lessing prefaced *Martha Quest* with the terrible words: "I am so tired of it, and also tired of the future before it comes." If Flannery O'Connor found release in dreams of violence, Oates's characters in acts of revolt or in conscious acts of submission, Lessing's women have become elegiac, resigned to a world where both violence and submission seem out of control.

VIII

JOURNEYS

If this life be not a real fight in which something is eternally gained for the universe by success, it is no better than a game of private theatricals from which one may withdraw at will. But it *feels* like a real fight.

William James,
The Principles of Psychology

Ernest Becker defined one of the most powerful early experiences as a sense of helplessness before the massiveness of creation, and claimed that virtually all human endeavor was an attempt to allay the anxiety it produced. That sense of universal inadequacy appears in new writing by women as a catalyst for change. Emerging when the novel is increasingly estranged from socioeconomic concerns, protest fiction by women personalizes an attack on the status quo by attacking the sexual drama of power and vulnerability. It uses the special female experience of passivity, dependency and susceptibility to launch an assault on inadequacy. As the passive hero flourishing in recent fiction is a sign of our anxiety about ourselves, the heroine, trapped and fighting, has become the living image of our will to overpower vulnerability.

One of the great weapons to emerge from the sexual revolution is a devastating she-wit. This humor develops absurdity as a female style, bringing the black comedian's sense of death and failure home as the joke on the facelift, the sexually bored husband. The bluntness of stand-up female comics like Phyllis Diller and Joan Rivers or the ironic satire of Lily Tomlin show how far our humorists have come from the hilariously inept Gracie Allens and Lucille Balls of another era. Women novelists show more deeply how wit can be a tool for self-discovery, for the unsentimental dissection of female lives. She-wit mocks the genres that embody old cultural ideals of femininity: the romance, the soap opera, the erotic love story. It transforms the weeping heroine into a joker. The woman writer as black comedian protests not through political statements but by an imaginative exposure of the female journey.

Can laughter light up woman's way to the American dream? Cynthia Buchanan's brilliant comic satire, *Maiden,* takes on America *as* a woman caught between the old romantic ideals and the new faith in sex. Her heroine, Fortune Dundy, is a thirty-year-old virgin with inflamed dreams and an impacted maidenhood, a hilarious model of America facing the loss of its innocence. This female Don Quixote is desperate for a man, but "she would not accept just any. This is America. . . . Does a person settle for second best? She waited for courtly love. An awful abstract, improbable lover! Someone on whom she might test her disdain." What are the prospects for an idealistic woman in the land that computerized intimacy? Will she find her man through Datamate, the computer service? Can she circle the five words that best describe her? Will she meet him at Dionysos West, the Southern California swingles apartment heaven?

Buchanan can make sex synonymous with the hyper language of a society in which the California superlative rules and the sheer mass of internalized pleas and promises exploits our massive insecurity. She has an absolute genius for making the loneliness of cruising singles thick and shattering as American noise. Her style is honed from glossies, TV specials, and disc-jockey late shows. It is as sharp and distinctively American as Flannery O'Connor's or Nathanael West's and wildly original at compressing essential messages into the clichés of a culture where people are taught to market themselves: "Sparkle was important. Employers know this." Charm schools advise "Be yourself." Fortune has tried this, but "was often punished for it."

Are the grossest national products people who embody sales myths instead of lives? America's leading escorts are often hawkers, walking self-help ads that entice by disparaging what you are. Fortune's superego is Miss America's escort, Bert Parks. "Bert was her different drummer. He pummeled her spirit without rationale . . . she was always being interviewed or questioned or complimented and then left to doubt the praise." Bert is the voice of her ambivalence, the dependency and self-hatred that keep her striving for success. Consider her at the edge of the Dionysos West pool:

> She sat down and dangled her feet in the water. The tops of her thighs could fool a lot of people. When I am sitting, Bert, my legs look more like they should . . . plumped out this way. Hah, there were so many little ways! So many little ruses! So many things the world would never know about her.
>
> All the women at the pool today did not look like secretaries. Some looked like princesses. Magic . . . with emphatic bodies. They lolled and sunned, undoing their straps. A person could, however, compete with them. If she kept her head about her. And used it.
>
> The thing to do, Bert, is be opposite. For example, sitting down now at the edge of the pool—that had been a blue-ribbon idea. A person could play toes with the water. She could turn her face sunward, stretch her throat up, up . . . pale sun goddess in mauve! Where did you come from pale, pale goddess?
>
> A horsey girl in a black tank suit swooped on the scene, rough as a lifeguard. She began to play in the shallow end, her broad brown shoulders wagging in the water. Ron and Buzz bobbed up next to this girl. The three ducked and splashed, wiping their hands over their streaming faces. The girl did not seem to be wearing a bit of makeup. Oh, she surely did think she was "something on a stick" for sure, that girl; swimming around, having close-ups with the boys . . . no, Big Brown, we do *not* think you are Esther Williams.
>
> An ambush occurred. A thrashing force pulled Fortune into the water by her ankles. She cried out. She clutched in panic at the pool's ledge. Away she went, underwater. She was drowning. She thrashed. From underneath, the water's surface glinted . . . a promised sky. She struggled to surface. She did, churning and gagging. Someone said, "Whoa! Sorry about that. Hold on there! You're okay. Put your feet down."
>
> Ron stood there, waist-deep. He was scooping Kleenex tissues out of the water. He handed her the pulp. "You lose this just now?"

"No! That is not mine!!!"

"You sure, honey? Take a look at the top of your swimsuit." She lost her mind. She tried to hunch down in the water. "Oh, this water does feel good!!! Oh yes—swim, swim, swim, just like a duck!"

She dogpaddled frantically toward the steps. But she saw beyond all question of a doubt . . . the rest of her stuffing followed her, disintegrating from the top of her swimsuit like pastel seaweed.

Ron called after her: "Man, I'm sorry. Didn't mean to be so rough." He waded over to the pool's edge and slopped a handful of Kleenex pulp onto the tile.

Do women who swing have it better? Fortune's roommate is her other self as Biscuit Besqueth, a divorcée in love again. They are two aspects of one woman: Fortune the romantic, idealizing men in her fantasies, Biscuit the woman who gets involved, gets hurt and even asks her lover to work her over. Her ex-husband, Skip, is a dentist whom she attacked with a barbecue knife during their marriage. Her new romance is peaceful enough until her man begins to wander. She brings him back by getting a breast lift and paints on her "new high riders," in Dayglo bodypaint, "Hey daddy, remember us?" Her lovers have made her cynical but effective. She thinks it's too late for love. Fortune is firmly convinced love is everything. She falls for Biscuit's ex-husband. Can romance resurrect the realist? Will Biscuit-Fortune rise again, whole?

The great sophistication of Buchanan's book comes from her sense of style, her use of clichés to mock the clichéd solutions of the hip chick and the romantic. She is lightly ironic about the destructive impact of intimacy on love. When Fortune is asked by Skip for a romantic motel weekend, she euphorically chooses the safari room. But before the crucial moment, Biscuit rushes in and shoots Skip dead. The chase that ends in the safari room seems to say that when you catch your man you kill romance, your capacity for love. Biscuit finds that love has only unleashed her anger. Fortune's virginity keeps her sweetness intact. Buchanan strikes the perfect ironic balance between celebration of love and utter realism about its aftermath. Her humor is brilliantly black in its implication that love must survive, if only as a hope. The price you may have to pay for optimism is lifelong virginity.

One of the finest and most ambitious recent novels, Buchanan's tour de force of unsentimental wit satirizes cultural ideals of love, sex, and romance and mocks both our fear of loneliness and our fear of each other. Her game and hapless maiden is no isolated sufferer, but America as a "Miss" in a grotesque dilemma.

Can wit soothe the individual heart and turn the soap opera of female life into a sit-com? Alix Kates Shulman's *Memoirs of an Ex–Prom Queen* attacks female life as a series of situational failures embedded in the basic female plot: first comes love, then comes marriage, here comes You with the baby carriage. What programs women for the story Shulman sees as both disastrous and inexorable?

The prime mover of female life may be irony. Shulman's heroine, Sasha Davis, discovers that the things that make women feel on top of the world are the ones that prevent them from getting there. She gets to be queen of her high school prom ("the triumphs of the rest of my life were bound to seem anticlimactic"). Envied by other girls, wanted by everyone, she has fulfilled all her desires. Dancing with the hero of the basketball team, her face proud and radiant, she hears the voice of male realism sizing up what so much female joy is worth: " 'If you don't get in tonight, friend, you never will!' "

"Coupled with good looks, surely knowledge is power," comments Shulman wryly. Can a woman manage both? At college, when her work is admired by her philosophy professor, Sasha immediately asks him, " 'Are you seducible?' " His intellectual judgement of her makes her anxious. Her sexual power over him makes her secure. She never develops the necessary intellectual tools of debate and argument to equip her to do anything with her work. Her sexual power over her professor does not bring her confidence or energy, but seems to decrease both. Happiness seems to make her exploitable.

Germaine Greer once said that she could not work when she was in love. What may be behind the peculiar paralysis of loving women is the often hypnotic sense of fullness and power that can flow from selflessness or the idea that female power is best expressed by giving oneself away. Sasha tells herself that she will let her boyfriend use her sexually because she is the omnipotent queen of the prom. She knows he will boast of his conquest, and she has no particular desire for him, but she tells herself that she is not too weak to say no, but too strong. Sexual doublethink kills choice, it undermines the capacity to identify emotions effectively, to look at them clearly and make necessary distinctions between self-expression and self-erasure.

Doublethink is a form of psychological bondage. Masochistic submission can be glorified as a form of self-transcendence; the woman who can take the most whipping is the one who is most sexually alive. Whether the perpetrator is a man with a whip in full porno-movie dress, or a society with inhumane values, the victim is so persistently denied the right to give her feelings their proper names that she loses the capacity to know what she feels. She becomes her confusion and ties herself up. She cannot climb out of the eternal disparity between what is and what is said to be.

Shulman has a fine eye for the extent to which a woman's joy can be the seed of unhappiness because it is expressed in ways in which seem designed to end badly. This disparity could be an ironist's dream. But for Shulman, the sitcom of female life loses its humor when women have to face the results of their bad choices. Sasha at thirty finds that the children her husband wanted drive him away, that the queen has become the pawn of crying babies. Shulman does not attempt to sustain the humor and the comic falls into the sad. But her excellent descriptive ironies sharpen Sasha's fall into a stark cautionary tale.

Can wit change the soul of woman? Can humor invade the recesses of the female psyche? In the new world of female absurdists, the masochist can mock herself and by doing so achieve some distance from the problem she knows too well to minimize. In this feminist fiction, when the worm turns, she does not turn on men, but on her own character. She admits her faults outright to stop anyone else from pointing them out.

The sick joke on oneself is a protective and hortatory device; it stops criticism and urges the heroine toward change.

Lois Gould's *Final Analysis* offers a psychoanalyzed woman who can do a lethal satire on herself in love with her ex-analyst.

> He is my ideal lover. He gives me what I really want. He knows all about me, and he's still there barely out of reach, allowing me to hurl myself at him all I want. And I'm not even paying him any-more—not in cash, anyway. Sometimes I imagine myself as a groupie, following this repulsive rock star who lets me get just close enough to rip his buttons off his clothes. He doesn't know my real name, but he'll certainly screw me one of these days. Which is just what I want—not for him to screw me, but for him to let me screw myself. So that's what I really love most: Hating me. The man who helps me do it is the only lover I understand.

Men seem like stage props for the S-M drama always in progress in a heroine's mind. The nameless heroine in Gould's novel affectionately absolves her lover of his many faults, satirizing him lightly as an amiable moocher with a tendency to show up hungry at other people's dinners. She forgives the magazine she works for when it defaults on promises it made to get her to take the job. She says, "I do five times as much work as anybody and never bitch . . . I stay until midnight if they ask me. I go home and cry so as not to bother them with my silly problems, and then I come back and work even harder so they'll feel sorry, and be nice, and let me go on playing with the big boys."

The loneliness of the masochist as comic heroine is embedded in the structure of Gould's novel, in which communication with her man is largely through letters he does not answer. The resolution between this man and woman is a comic accommodation that you might call love. He offers her an out from her job by lending her his summer house without himself for the winter so she can write a novel. The separation makes the affair a success. Distance makes the heart grow fonder than proximity could permit.

When detachment ends, when involvements tighten, humor gives way under the sheer weight of released pain. Humor is an effective weapon in novels of courtship or superficial affairs. Whether the joke is on society, on female situations or on the psychology of women, the best female absurdists keep their humor going by keeping their characters away from deeper, binding involvements with men and with children. In contrast protest novels of intimacy vary from a female literature of disgust to sexual fables; but they all explore what paralyzes women with a directness fully charged with pain and fear. They protest through dramatic enlargement of the odds against female happiness.

Lois Gould's *Such Good Friends* showed there could be a female literature of disgust. Gould opens up female sexual anger as a subject, and anatomizes the price a woman can pay for marriage. Julie Messinger discovers, while her husband lies comatose and dying, a coded diary of his infidelities. Positions used, orgasms experienced, dates, initials of mistresses are listed by the scorekeeping husband who rejected her sexually. Gould compresses Julie's humiliation into a slashing, quipping style that gives her heroine's anger a concentrated power. Not a lament-against-a-husband, the novel is a revelation of a frustration so great it makes her see herself through her husband's angry eyes.

In Julie's sexual fantasies her husband controls and kills her pleasure with someone else:

> "You can't keep rubbing her goddam tits all night. I know she likes it. . . . You kidding, it's the only thing she likes, but whose party is this? I mean, it's my house. And you can get your goddam finger out of her snatch, too, for Chrissake. Can't you just get in and fuck her? You wanna make *her* come? You're kidding, why? Five little quivers and she zonks out for the weekend, believe me. Besides she'd rather do it herself when you're through. . . . You oughta see her at it sometime, it's fantastic. She used to lock herself in the john, took in the Times Drama section so I'd think she was just in for a crap, the stupid cunt. I actually watched through the keyhole one time and there she was on the floor. . . ."

This is less an indictment of the controlling male than a revelation of being controlled as the pervasive experience of a woman's life. The man who calls the love-plays is Julie's own creation. Masochism *is* the spellbound love that originates in a sense of inadequacy and dependency so great one yearns for a redeemer, attributes to him almost superhuman rights and powers. It is a form of idealism subverted to complaisance. Love-struck in happiness or wretched victims in bondage, masochistic women remain in need of a figure who represents authority. Julie is locked into subservience by self-hatred and need, by the unquestioning acceptance of her husband's right to use and humiliate her.

Self-haters, needers, sufferers are made, not born. Julie's husband is an incarnation of her suffering, the voice of her lifelong habituation to second place. In a few adroitly drawn scenes of adolescence, Gould sketches the situation of a girl whose mother is "perfect." "My mother was one of those who simply do not carry packages, except perhaps a small hostess gift that looked so beautiful it was a shame to open it. For a small service charge, she moved through womanhood in clean, eight-button gloves." She berates her daughter for necking at a party, insisting, " 'I have never felt a need, not the slightest momentary impulse to degrade myself for some disgusting physical urge. And I doubt very much if I've missed anything worthwhile.' "

This kind of humiliating tirade is designed to make sure that Julie becomes no competition to her and to erode Julie's confidence in her own impulses. Self-hatred is the other side of Julie's fear of competing with the slim, chic, perfect mother. Julie as a college girl is obese, unkempt. As a woman she is slender and dresses well. But she repeats her childhood humiliations by pleasing men who abuse her. She performs fellatio on men who do nothing for her in return, and mocks herself: "Now ladies and gents, the contract demands that Julie the Sword Swallower perform her world-famous teething ring trick."

How Julie is with men reflects how she was with her

mother. Age brings an eroticization of childhood feelings of worthlessness and submission. One of the women writers to take on the destructive bond between mothers and daughters, Gould probes situations in which a daughter's pleasure has to be stolen from a mother's ill regard.

> Once at the age of five Julie had seated herself gravely before the altar of her mother's dressing table. She would perform the rites and change herself into the beautiful Mommy. . . . And then suddenly footsteps clicking in the hall, and she had jumped in terror, knocking over a tray of perfumes. She was not punished because she'd gotten sick from crying. But she would never forget the overwhelming scent in that room. The essence of her mother's beauty which she had only wanted . . . to steal. What had happened was an official warning.

Julie's lifelong depression wards off her mother's envy. She does not outdo her mother in happiness, and she insures a life of failure by staying with a man who makes her feel like one.

Gould exposes the extent to which a woman's sense of security can become tied up with being left out, stewing in inexpressible feelings, playing competitive games to lose. Julie's emotions become centered on the experience of feeling rejected and controlled. In her marriage she keeps alive a vision of herself as being in the way, unattractive, having no right to anything herself. In marrying a man who sees her and his children like this, she virtually marries her mother.

Is there any alternative to spending your life stewing in impotent anger? Joan Didion's women deal with their pain by going dead to it. In her beautifully crafted novels, women drift in a virtually unshakable depression wherever the strongest current takes them. These women mean no harm, no evil, but are always there when it hits their loved ones. They are fragile heroines who just happen, in two of Didion's novels, to be the "loving" women who help their dear men die.

After her husband shoots one of her lovers, Lily, the heroine of *Run River,* sits down on her needlepoint chair while her husband goes upstairs to kill himself. She thinks:

> They had been a particular kind of people, their virtues called up by a particular situation, their particular flaws waiting there through all those years, unperceived, unsuspected, glimpsed only cloudily by one or two in each generation . . . by a blue-eyed boy who was at sixteen the best shot in the county and who, when there was nothing left to shoot rode out one day and shot his brother, an accident. It had been above all a history of accidents.

What can people get from seeing themselves as the passive victims of chance? Didion's heroines get off the hook of responsibility, and find the ignorance that is bliss. Didion is among the few women writers who see fragmentation as a force for survival, who applies anarchic, disintegrative themes to the lives of women. Her heroines split what they feel from what they do so radically that whatever happens seems like a surprise. They cannot put together what made them what they are. Maria in *Play It as It Lays,* Didion's most dismantled character, is developed through a series of disconnected memories that she will not permit to cohere. Her nerves have no synapses; no leaps take place to transmit the locus of pain to the mind. "What makes Iago evil?" she says. "Some people ask. I never do." Her dullness is a form of relief, her numbness a saving grace. Who gave her this Novocain? "My mother's yearning suffused our life like nerve gas," remembers Maria. "Cross the ocean in a silver plane, she would croon to herself and mean it, see the jungle when it's wet with rain. The three of us driving down to Vegas in the pickup and then driving home again in the clear night, a hundred miles down and a hundred back and nobody on the highway either way, just the snakes stretched on the warm asphalt and my mother with a wilted gardenia in her hair." Evil as "nerve gas" diffuse as a dream, evil as a "climate" suffuses Maria's life.

Maria feels she is "confirming a nightmare" when she gets a hypnotist's circular in the mail and reads, "YOUR WORRIES MAY DATE FROM WHEN YOU WERE A BABY. IN YOUR MOTHER'S WOMB." She recognizes her mother as a damaging force in her life but cannot face the fact that she was in any way against her. Her anger paralyzes her, produces no fury but only "nothingness." Depressed, Maria travels to Hoover Dam and begins to "feel the pressure of Hoover Dam here on the desert. . . . All that day she was faint with vertigo, sunk in a world where the great power grids converged." The gigantic dam is an image for her anger. The closer her anger comes to consciousness, the more feeble she feels.

Anger is available to Maria as a negative space, a void. She expresses it through withholding. Quickly fulfilling her mother's dream life, Maria becomes a successful model, actress and producer's wife and saves enough money to send her mother around the world in a silver plane. But instead she lends the money to a man she barely likes. She cannot bring herself to give her mother what her mother wanted; nor can she enjoy the life her mother would have loved to live. Looking miserably on her easy winnings, she observes, "I was holding all the aces, but what was the game?"

What Maria absorbs in infancy is her willessness, her paralysis of self, her habit of playing roles someone else assigns. Long after her mother is dead, she goes on woodenly saying "I love you" to her husband and "you make me happy" to her lover, who makes her anything but happy. The experience of destructive love is so primal it becomes the eternal reality. Love and hate are so mingled she can't tell one from the other.

Maria's anger at having complied her way through life is finally expressed through her love for Kate, her neurally damaged four-year-old daughter who responds to no one. Maria loves her as a sister in neural collapse. She has fantasies for Kate that are as unrealistic as her mother's were for her; she wants Kate to save *her.* She wants to be Kate. Kate has the absolute freedom of someone without attachment to anyone. Maria plans to build her life with Kate on their mutual inability to be reached.

Maria sees pure emptiness as her only available relief. She hopes for the invulnerability of the burnt-out case who has been hurt too much to be hurt again. Pregnant by her lover, about to divorce her husband, "eager," she says, to have a child, Maria nevertheless gives in to her husband's pressure to get an abortion. While she blames him for it, she clearly hopes to find some release through eliminating the possibility of a healthy child. She tells herself that "she would do what he wanted, she would do this one last thing, and then they would never be able to touch her again."

How unreachable Maria becomes is underlined in one of Didion's best scenes. In a nondescript motel room Maria is aware of nothing but affection for her friend, the homosexual BZ. He puts a bottle of barbiturates on the bed between them, tells Maria he is going to take them and does, while she holds his hand and drowses naturally beside him. She sees her complicity in his death as a gesture of kindness. Like Lily, she is all love for the man she lets die. BZ is also a kindred spirit, a more aggressive self who does literally what she does spiritually. Maria accepts a life with men in which she is a victim of their need and indifference. She plays so many bad scenes that they begin to seem rather alike, uniformly uninteresting, and reaches an erosion of herself that permits her to be free of caring. In a sanatorium like Kate's at the end of the novel, she throws the *I Ching,* but does not bother to read what it says. She thinks of her life as a gamble; but in fact she had made it certain and secure as paralysis.

Can you control vulnerability by becoming an aggressor? Is life less painful if you act on your anger? The determination not to be a victim makes some women identify with the attacker. One of the newest themes in fiction is the sexuality of women who have eroticized their own aggression. Judith Rossner's *Looking for Mr. Goodbar* is a richly sensitive, holistic novel of a woman alone in New York, where a female ideal is the girl in the singles bar. She loves her job; she has no particular wish to marry; she picks up whomever she pleases. She looks like she's gotten out of the old female traps. But Rossner's novel begins with this woman as a murder victim who was destined to be one. Does the apparently aggressive woman work out a compromise with life that kills? Is aggression the other side of passivity and suffering?

Rossner builds a past for the new heroine who seems so historyless before her beer. Terry Dunn was meant to be lonely. At four she was partially paralyzed by polio, at eleven immobilized for a year in a body cast after surgery for curvature of the spine that had gone unnoticed, and her back has been broken still more by her parents' lack of involvement. Her mother is intensely concerned with her older brother who died in the war, her father with her beautiful older sister, and both with a younger sister, a funny lively tomboy who grows up to be a mother of three. Terry knows she will never have the ease of the beauty who glides through life acquiring and discarding husbands, or the physical sufficiency of the girl who is simply at home in her body and likes it for its functional efficiency in playing baseball or having babies. She equates her inner life with the hideous scar on her back, which she keeps covered. She is eager to please, grateful for whatever small crumb of affection is thrown to her, and willing to work hard to keep crumbs coming.

What Terry takes from her past is the habit of paralyzing her resentment. At college she falls in love with a teacher who likes her work. For his moderately kind words, she becomes his underpaid typist and sexual convenience. After four years of being serviced, he discards her. She drifts into a series of initially trustful, painful contacts, picked up by men who use her and shed her with varying degrees of politeness. Her bad experiences gradually break the cast on her anger. She expresses her resentment through an increasing identification with the men who exploit her. She sees herself as one of them.

Terry gravitates toward gross, criminal types, excited by men who seem as willing to knife her as to make love to her. She is turned on by feeling intellectually superior to them; her initial kick comes from seeing the power given her by their sexual desire for her. She contemptuously lets them buy her a drink. Sometimes the ensuing sex makes her ecstatic and sometimes she's left frustrated and discarded. She takes the chance of being beaten or abused to see herself as the woman whose machismo is equal to it all.

Rossner explores the extent to which hanging on to relations which center on aggression and anger can reflect hanging onto masochistic pain. When the obvious husband turns up for Terry, she's revolted. This Irish Catholic lawyer is charmed by her, enjoys listening to her talk about her work, shares her enthusiasms. Her tenderness, reserved for the children she teaches, comes through in their conversations. But his goodness arouses every fear of involvement she has. He revives her sense of paralysis as a child; he paralyses her in bed because of his respectability. His ability to control and integrate his own life arouses her fear of being controlled as she was as a child. It is his persistent offer of marriage that pushes her toward the extreme provocativeness that gets her killed. She picks up a man like the others she has picked up—a drifter with a prison record. She tells him to leave after he makes her come. When this has been done to her she has left quietly, if drearily. But this man chokes and beats her to death and comes in her dying body.

In the new world of female aggression the earth is a starkly simple place inhabited only by lions and lambs. To avoid being devoured, the lamb must go through a more radical transformation than imitating the aggressive male. She can reject womanhood as a weakness. Louis Gould's *A Sea Change* is a fable for the new jungle warfare of sex, a work of a radical, feminist imagination in which a woman sheds her vulnerability by becoming a man.

Woman as incorporator, the engulfing vortex of the hurricane, the vagina dentata, and man as the thrusting rapist, the knife, the gun are the yin and yang of sexual anger. Gould reconciles the contradiction between male and female by making the common source of all human relationships the will to power. The characters of this novel are additions to a female literature of disgust. They are not so much people as personified human drives.

Jessie, a beautiful model, a mother, a wife, in the opening scene kneels tied up by B.G., a black gunman who is robbing her house. The robber's gun triggers her anger. Tied up, she first sees her lifelong passivity. Poked by B.G.'s gun, she realizes her anger toward her husband, Roy. The death of her affection and need for him frees her from fear of him. His rages no longer have any impact on her. The man who seemed controlling, in charge of her life is reduced to ineffectuality as she rents a house on the tip of an island to which he can only commute on weekends, as her silences lengthen and nullify him. He disappears on a business trip, leaving her with their daughters. She invites her old friend Kate to visit and become Kate's lover. As hurricane Minerva, which will destroy most of the island, hits, she grows stronger, drawing the force of the storm into herself. She domineers, she snaps at Kate. When a coastguardsman comes to rescue them, she gets him drunk and asks him to rape her. Incorporating his brutality, his strength, she is magically transformed into a man.

Gould as a fabulist builds the drama of sexual power and vulnerability into an S-M bond of mythic proportions. She draws her image of the sadistic male from the street, her image of the victim from the middle-class woman who is the target of the rapist, the robber. When Jessie turns into B.G. her first act is to rape Kate: "Kate only whimpered a little when B.G. said what he was going to do. . . . he reached under the pillow for the gun. He slid it slowly down until it reached her groin and then pressed the barrel against her, there, hard. Inside, he said softly. All the way in. That was when she started whimpering again. No. But all he had to do was nudge her a little with it, and she went down the way he wanted her to. He stood for a moment astride her body, just looking down at her." He makes the coastguardsman watch while he forces Kate to orgasm on the gun.

> She had never been forced to confront her visceral response to that kind of power. He had come to make her do that; make her body admit what they both knew it felt. And he had succeeded. That was much worse than brutality or humiliation. . . .
>
> As for the weapon, she understood that too. His power had less to do with the gun he was using than with her reaction to it. Her repulsion, her rage, her hostility. He had made her see that they were all his weapons, not hers. He had used them all against her.

By force of his force, B.G. wins Kate's love for life. As the hurricane diminishes, B.G. takes advantage of the debris, confusion and wreckage to leave for a still more remote island where he hopes to live as a man with his submissive woman, Kate.

What makes Gould's resolution of the novel hit with the force of a thirty-foot wave is that B.G. is grim, flat, as frightening as death. Jessie has paid the price of tenderness, love, joyousness in order to live only in and for the continual spectacle of power. Neither B.G. nor the gun "comes" in the rape scene; the high in sex comes from being the victor. Gould does not make any simplistic attempt to claim that sexual rage disappears between female lovers. As an explorer of intimacy in a world where the drives for power and for sex are increasingly synonymous, Gould binds male, female, homosexual and heterosexual into a common wish to be in control.

The transsexual is a product of our time, the creation of a surgical technology that has only existed for fifteen years. The transsexual hero in this novel is the product of the fact of continuing female vulnerability, of rejection of everything female as irremediably weak, of fear that in mastering others we may become grotesques. The power of Gould's novel comes from its daring, its declaration, in the face of the harshest possible vision of human relations, that even this price may be worth paying. Gould remains open to the deepest ambivalences in the female condition. Jessie's transformation into B.G. is no simple success story, but a sophisticated recognition that power can be both destructive and essential.

It is precisely this kind of sophisticated rawness that gives protest fiction by women its unique force. In most protest novels the heroine's standard devices for controlling vulnerability are simply not adequate to controlling the kind of emotion involved. The failure to develop adequate defenses may be a torment in life but is a remarkable asset in fiction by women, adding to its emotional depth. (Male characters in comparable books by men are often so effectively defended they seem incapable of sustaining any strong emotion.) As a result these novels by women do reflect the charge of feeling often present in our lives, and the inadequacy of our present solutions to our problems.

The perception of life as a struggle for survival between

lions and lambs is a mark of how scarce the better emotions have become. Women *are* willing to kill themselves to avoid getting slaughtered. It is not surprising that in the novels which show our revolt against victimization most blatantly, the heroine identifies with the most extreme macho stereotype—the criminal, the rapist, the murderer. The worst male characteristics are idealized as preferable to any they might possess, to any old or new female trait. It is better to be a lion than a lamb, better to devour than be devoured. But it would be best of all not to live in a jungle. What makes us see ourselves as only murderers or victims? Why do we judge so harshly both our aggressiveness and our weakness?

Protest fiction makes clear how self-hatred, impotence and fury blind us to humane and effective alternatives. Women characters swing between an unreal sense of omnipotence and an equally unreal infantile submissiveness. They repeatedly overestimate the destructive potential of their anger. Like small children and the insane, they believe their evil wishes will come true. Rage is a hurricane that destroys the eastern seaboard; it erupts into murder; it can be projected, like Martha Quest's anger in *The Four-Gated City*, as an apocalyptic third world war. But rage rises to such dimensions in proportion to how much it is suppressed. Blocked by cultural sanctions, by the dependent's fear of losing the good will of whoever is hated and needed, it grows, becoming still more frightening to face. The more fearful it becomes the more it is expressed disguised as love (Didion), as sex (Rossner and Gould), the more it destroys the soul (Gould). By such routes aggression becomes a twisted force in our lives, not one for productive change.

The excitement of protest fiction comes from its courageous exploration of our emotional lives. It brings an almost magic realism to women's attempt to use what means are at hand to resist the pain they see as inherent in their condition. It reflects an honesty about what we are not that is itself a force for mastery. By authoritatively exploring the female journey, protest fiction repossesses it from its old entropies—depression, passivity, and hopelessness—and extends it toward the future.

The heroine as joker or sad sack is the image of our attempt to live happier lives. The heroines of protest novels are blow-ups, truths enlarged to unavoidable dimensions. As a feminist I am a social optimist. I do not believe the assertiveness of women will founder on absurd male models or be dissipated in vindictive, retaliatory dreams. What will change our lives in ways that are both humane and female is a recognition of the internal struggles that erode our energies. Our protest novelists radicalize us by *not* being ideologues. They expose to the light what is darkest in us, what keeps us in darkness, what takes us down so many bright roads which turn out blind.

IX

THOMAS PYNCHON AND WESTERN MAN

Happy is he who knows the causes of things.

Virgil, *Georgics*

Thomas Pynchon is a private man. He did not show up to accept the National Book Award he won for *Gravity's Rainbow*. He quietly acquired the O'Henry Prize for short fiction, the William Faulkner Foundation First Novel Award for *V.*, and the Rosenthal Foundation Award for Fiction. When in 1975 the American Academy of Arts and Letters awarded him the Howells Medal for the most distinguished fiction produced in the preceding five years, he would not accept it. He is said to have refused (in what would be an unusual literary event in this decade of self-promotion) to have his picture taken for the cover of *Time*.

The most daring and ambitious American writer alive today, Pynchon lives out the myth of the artist as a man apart. He is the Americal El Greco whose canvases seem lit from another world. Pynchon makes us see by the lights of heaven and hell how modern history has altered people, how much the dream of our age has become the dream of conquering vulnerability. At twenty-five Pynchon dared to say that what his generation required was salvation from death *and* life. This ironic moralist put life together as a diabolic pact in which you could trade your soul for insurance against hell on earth. His first novel, *V.*, showed the way to eternal experience without anger, pain or fear. Published in 1963, it was set in 1955 because the cold war was an unbeatable image for the stand-off between Eros and Thanatos in suburban marriages, in New York games of musical blankets, for the deadlock whose linear representation was the symmetrical letter V. Pynchon saw the freeze as an emotional necessity. He wrote about people who knew that love could not diminish suffering because it was love that produced half the anguish there was. He knew that what the world needed was not another Christ, but an end to the daily passion play.

Pynchon's symbol for human salvation was not the cross, but the partridge in the pear tree: the bird lives off the pears; his droppings fertilize the tree so it can make more pears. The bird makes more droppings. This is an image of nature as a Newtonian motion machine powered by crap. Salvation is mechanical symbiosis between *people.* The Prime Mover shows you how to keep it going without upsetting the bird! Pynchon's Christmas present to his audience was the God, the messiah who was a birdbrain machine.

Technology is commonly blamed as the source of all our woes, our short-circuited relations, our IBM-ized lives. But many people do not fear machines. They envy them. Pynchon loves and hates his messiah machine, Benny Profane, a human yo-yo whose nightmare is that his "clock-heart" and "sponge" brain will be disassembled on the rubble-strewn streets, but whose grace is his ability to be a perpetual-motion man who rolls on too fast to lose his heart or let anyone touch the controls of his mind. The Profane Christ is the one who won't get crucified.

Profane's world is no vale of tears. His nativity is on Christmas Eve in the Sailor's Grave Bar, the hip world where every man is a drunken sailor and women are interchangeable quick lays. Everyone's waiting for Suck Hour, the moment when Chow Down calls the sailors to custom beer taps made of foam rubber in the shape of large breasts. There are seven taps and an average of 250 sailors diving to be given suck by a beer-breast. There's very little nourishment in Pynchon's world; his wise man controls his thirsts. Profane does not really want to turn on anything, even a beer tap. He wants a woman who will not love him but will be a really self-contained machine. "Any problems with her you could look up in a maintenance manual. Remove and replace was all." He gets an erection thinking about the sex money can buy while reading the want ads, and notices that his erection traces a dent in the *Times.* But he waits until it subsides so he can choose the agency where it comes to rest. He wants the least exciting job. He has the peace that passeth understanding.

What virgin bore, what holy spirit sired Profane? It took the male and female forces of modern history three generations to produce him. Profane's inanimateness is the twice-transformed version of a Victorian belief in order. Sidney Stencil is Pynchon's favorite Grandfather, a member of the British Foreign Office whose life's work was the prevention of chaotic, violent situations. He was called "Soft-Shoe Sidney" because of his mania for performance and teamwork and his belief in the dance of civilization. He thought he could choreograph violence, align hostile forces in a dance. Out of a flirtation with the mysterious woman V. he had a son, to whom he left his notebooks. But Herbert Stencil cannot learn his father's lesson.

Are sons dimmer than their fathers? Herbert is only a shadow of "modern man in search of an identity," a cause. His intellectual quest moves him only from apathy and "inertness to—if not vitality then at least activity." Abandoned by his mother at his birth in 1901, depleted by living through two wars, he wants to put together the pieces of his personal disintegration. What his father hoped to save was civilization. What Herbert hopes to save is himself. Sidney wanted to find solutions; Herbert wants to extend his problem because his quest keeps him going. Pynchon believes there was a shift in the goals and scope of thought from the driving, syllogistic means to a solution to an obsessional model of the problem. Herbert's reasoning is like a Wittgensteinian proposition in which each term is a model of his reality. His mind is a series of dead ends. For Herbert life is possible only as the romantic pursuit of an unattainable meaning, an unattainable woman. The purpose of the hunt is to hunt forever.

If you find your Venus, you find death. Hugh Godolphin is Grandfather Romance, the Victorian explorer who believed in his immortal soul and the British burden. He lost his faith in Vheissu, an "outland" arousing sexual appetites symbolized by iridescent spider monkeys who are all cling, a place like the skin of a "tattooed savage" whose wild beauty and "gaudy riot of pattern and color" would make you fall hopelessly in love. But her beauty would get between you and "whatever it was in her that you thought you loved," and in a matter of days "it would get so bad that you would begin praying that whatever god you knew of would send some leprosy to her. To flay that tattooing to a heap of red purple and green debris, leave the veins and ligaments raw and quivering and open. . . ." Godolphin escaped to the South Pole, hoping to find his soul again. But digging a hole to plant the British flag, he uncovered a dead spider monkey and found a network of tunnels leading from Vheissu to the pole. The heat of sex is bound up with the ice of death. One leads to the other because, in this novel, intimacy kills.

Hugh Godolphin's son, Evan, made hedonism a religion, avoided intimacy and never thought that much about his soul. But his plane was shot down in the Great War and his face smashed to a pulp. It was repaired perfectly with paraffin which, his surgeon says, will melt him back to a blob in six months. He put his faith in a body subject to age and accident, forgot his soul and became a freak. In Pynchon's vision of history as the change from fathers to sons, the mind that is an escape from chaos becomes an involuted trap; sex kills romance, and war perverts sex. All you can be sure of is your paralysis in vulnerability.

Father Evan and Father Herbert sired Profane in the belief in reality as a phenomenological problem. Their generation produced the developments in plastic surgery that put people back on the skin market, devised the mechanical solutions that kept them going in the hope they would find meaning. This is the generation whose veneration of the qualities of wholeness, durability and progress, whose emphasis on thought as logic, as concrete models of experience, surfaced in the works which hover through Pynchon's imagination of the period: Wittgenstein's *Tractatus,* Husserl's *Ideas* and Heidegger's *Being and Time*. This is the generation whose effort to control experience only produced the profanes who will not try to shape anything, who do not look for meaning in the illusion of meaning but in embracing their historylessness, their replaceable clock-hearts, as values.

The history of male striving can be written in excrement, as Norman Brown implied. Pynchon wrote it in his wacky sewer scenes where evil is the excremental tunnels of New York's sewers, the world where everything is simplified to human waste and three generations merge. The old priest Father Fairing who went into the world to preach the word of God wound up seeing people as rats trying to get sanctified. At least he believed the rats had souls! Middle-aged Herbert Stencil hopes to wrest personal meaning from the surge of crap around him. Young Profane wades through the remains of their hopes daily, on the sewer patrol just to earn the money for food. He makes the directionless flow of crap his life.

V. is female serenity, the clean, eternal balance of emotional control. She absorbs the shock of war, of male striving, as an erotic curio and returns it when, as mother, she abandons, as lover she murders and as protectress she corrupts. She is the indestructible woman who sees herself as an objet d'art, who mutilates her body to become one with golden feet and a glass eye. She is always young, always fascinatingly beautiful. Stencil dreams of her ecstatically as a young machine: at "age 76, skin radiant with the bloom of some new plastic, both eyes glass, but now containing photoelectric cells connected by silver electrodes to optic nerves. . . . Perhaps even a complex system of pressure transducers located in a marvelous vagina of polyethylene, all leading to a single silver cable which fed pleasure voltages direct to the correct register of the digital machine in her skull." She is Profane's woman, the girl who lost her virginity to the gear shift of her MG, whose great love is her car or its human equivalent, Profane. V. is a self-contained autoerotic machine. V. is the crucial pivot, the profane fulcrum on which you can survive forever.

The degree to which men and women want each other to be ever-ready erotic tools, needing neither tenderness nor love, is one sign of sexual hate. Pynchon is saying that men control their destructiveness through Profane-like passivity and disengagement; that women conquer their vulnerability to men, life and death by becoming virtual automatons who cannot feel a thing. "Keep cool, but care," someone advises. The only way to contain your destructiveness is to deadlock the two, to be the partridge and pear tree locked in endless, mechanistic, profane life, forever content.

"O trees of life, when will your winter come?" asked Rilke in the *Duino Elegies.* When V. topples, when destructiveness is not deadlocked against any control, when controls become forces for death, when the cold war between life and death heats into open battle, life may seem to lose its generative power. Pynchon took up Rilke's insistence that life be opened toward death. But Pynchon out-Rilked Rilke in unleashing death on life. The last chapter of *V.* ends with Benny Profane and a pickup who owns seventy-two pairs of Bermuda shorts running down a street toward the sea. However Pynchon added an epilogue in which his parting shot was not the warm run of sex, but murder. In the Epilogue, dated 1919, he leaves you with Evan Godolphin, the smashed-face freak who may have engineered Sidney Stencil's shipwreck and drowning, watching Stencil's ship sail off towards its doom.

Pynchon could not take *his* eye off civilization and the soul's death. *V.* is Pynchon's *Sonnets to Orpheus,* a book of life written with intelligent compassion by a man who wants to survive the touch of death. The holistic impulse to pattern personal aggression and the disorder of modern history, and the disintegrative love of chaos, balance in *V.* like those ancient antagonists life and death. But Pynchon rejected the balance and became, like Rilke, an elegist. He methodically went into the breeding ground of emotion, unfroze the spider monkey that is all clutch, and staged an open war between life and death. Death won. Pynchon descended towards a new vantage point, going from an ironic modern heaven toward an ironic hell. He invented himself as the Devil, the fantasist whose rainbow has its origin in gravity, the spirit of the down, of depression. *Gravity's Rainbow* is death's fantasy that life exists.

World War II is an irresistible image for Death's primrose path of heroes and villains who kill each other off. Pynchon's psychopolitical fantasy of war, for all its stunning historical detail, is an apolitical circus in which national differences do not matter and allies and enemies are more dangerous to themselves than to each other. The combat unit for Pynchon is the whole Age of Aquarius encapsulated in the microcosm PISCES—Psychological Intelligence Schemes for Expediting Surrender. In that psychological warfare unit, it is never clear whose surrender is being plotted because everyone is busy devouring everyone else.

Why is the world so full of hate? How did death beat out life? In his American way, Pynchon embeds his question in a western. The fastest gun in London is Tyrone Slothrop, an American officer who is the ultimate lady-killer. The places where he has gone to bed with his pickups are the exact spots where the V2 rocket falls. The psychological warfare unit knows this because Slothrop, who is a member of it, keeps a map, charting with gold stars the places where he has scored. Yes, part of the problem is that he is exactly the sort of man who would do this. Roger Mexico, a statistician, charts the bomb sites. His map and Slothrop's are identical. Is it the bomb that excites Slothrop or sex that draws destructiveness on the girls? Which came first, the bloodsurge or the blood-bath? Will anyone stop the deadliest gun in London?

America's good guys are the practical engineers who claim to have all the answers. Pynchon makes his emotional points through their "practical" expertise. Pointsman, his Pavlovian, believes there is a point, a particular switch in the brain that turns on sex or death. If only he can find the mystery stimulus that controls the switch, he could turn off death and win the Nobel Prize! He could end the war between the sexes! Pointsman salivates while devising ways of checking out Slothrop's erections through a system of spies, seductresses and voyeurs, longing like the creep he is to kill Slothrop's one enthusiasm and eventually to castrate him so as to measure every drop of love. Slothrop begins to suspect that his penis is no longer his own. Paranoia rules as Pointsman's stimuli leave Slothrop less and less able to tell pleasure from pain, dominance from submission. "Paranoia even Go-ya couldn't draw-ya!" sings Pynchon. But Pynchon drew it in this fantasy: We are all dead and have been for years. Perhaps the Devil is tricking us into believing we are alive.

What looks like the creative intelligence of a Pointsman is the work of a man whose models of human reality as off/on switches emit the gases of the grave. Pynchon's harshness toward Pointsman is the mark of his total rejection of his own belief in the right tactic, the balance point that can prevent human relations from toppling into death. What Pynchon now hates is the mechanistic, partridge-and-pear-tree vision of human nature, the will to find the point outside in space—in spaciness!—from which you can move the earth, or keep the life-cycle going.

Mathematics increasingly allows for pointlessness, contingency, probability. Pynchon's anti-Pointsman is Roger Mexico, who tells the Pavlovian that there is no explanation for the identical graphs for Slothrop's pickups and the bomb sites. "Bombs are not dogs. No link. No memory. No conditioning." Mexico seems to be happy with the discrete, chancy droppings of the bomb. Gödel's proof showed the existence of mathematical contingency, of unprovable assumptions in mathematical systems, in effect incorporating chance by institutionalizing it. Mexico is Chance, Inc. He is desperately clinging to meaninglessness to avoid the obvious fact in his life: his intense sexual passion for an unloving woman will be the death of him. Contingency, probability are ways of clouding what Pynchon ironically reports as a fact: sex and death are the same; slaughter is a certainty. As Rilke wrote to a friend, "The future is stationary, dear Herr Kappus, but we are moving in infinite space."

Points and pointlessness, meaning and meaninglessness are opposite sides of the same delusion, diversions into the traps of control or chance and away from the fact that there is a point to life and it is uncontrollable. For Pynchon only physicists give clear unequivocal statements that Death has his undisputed hegemony in the universe, that life moves from order to disintegration, from differentiated structures to dispersed, undifferentiated matter according to the second law of thermodynamics. Pynchon's law of human entropy orchestrates the life of the nation, the couple, the family, the individual into a symphony of death centuries in the unrolling, its pattern inaudible to any one listener because a lifetime unfolds only the most miniscule movement, tricked out as the song of life.

What could look more like life than sex with an irresistible blonde who makes you feel like a leading man? Katje, sent by Pointsman to check out Slothrop's performance, makes him feel like a hero. He suspects she's out for more than she lets on, but he is, too. He hopes she knows the secret that will unlock his humanity and make him feel some emotion. "His face above her unmoved, full of careful technique—is it for her? His desperate hope, She will move him, she will not be mounted by a plastic shell . . . her breathing has grown more hoarse, over a threshold into sound . . . thinking she might be close to coming he reaches a hand into her hair, tries to still her head, needing to see her face; there is suddenly a struggle, vicious and real—she will not surrender her face—and out of nowhere she does begin to come and so does Slothrop." The hope for a woman who will connect, who will be your connection to life, is betrayed by the recognition that the woman is as plastic as you are. What looks like the movements of love is really the dance of death.

Detached sex is depressing, but involved sex may be death. Pynchon sees women who can connect you to the center of yourself, whose intimacy connects you to their center, who are in touch with their feelings, as only in touch with evil. Greta Erdmann, an actress whose *life* is an expressionist S-M movie, is Pynchon's vision of woman as lover, mother—a "total" woman who demands to be gang-banged by the entire cast dressed in monster costumes after a filming, whose daughter is fathered by one of these beasts, who commits a series of child-murders, who raises her daughter for S-M incest with her and eventually murders her. Slothrop only dimly realizes in his affair with her that he craves cruelty too and is not the good guy he thought. She looks into a mirror one day and ecstatically sees the face of the Devil. But Slothrop dreams of her as the Earth Mother, the genetrix at the bottom of an industry-poisoned river, her womb breeding all manner of monsters into the tainted water. Who fathered the mutants?

The Devil of male industriousness, the polluted orgasms of industry, the male mind that creates structures, forms, controls that kill life, is the putative Father. Pynchon's Devil is a formalist; his evil is his ability to rape nature with elegance, with all the classiness of Thomas Pynchon's symmetrical alignments. The Devil who did in Slothrop was Kekulé, the father of synthetic chemistry, who dreamed of *his* baby, the benzene ring, with an X-ray vision that revealed all the hidden structures of life. He dreamed of it as the great Serpent who surrounds the world, its tail in its mouth, symbolizing the world as a closed thing, cyclical, eternally returning, inviolate. But he was only looking for the weak link, the vulnerable point where he could strike. His vision began the system that produced the plastic man Slothrop, that substituted for the eternal return the movement from death to death transfigured, the development of synthetic polymers whose origin and structure reflect gravity. In human terms male destructiveness is expressed in Blicero's homosexual sadism toward young Gottfried, the boy who, looking innocently at his "master," a Nazi engineer, hears, "Can you feel in your body how strongly I have infected you with my dying? Fathers are the carriers of the virus of Death and sons are the infected."

Is the history of death the history of parental love? Slothrop is granted a buffoon's revelation of creation when he throws up in a barrooom toilet and drops his harmonica into his slop. He dives in after it, only to get heaped with the excrement of others and flushed into a wasteland where he sees the souls of babies waiting to be born. They look like the remains of basket cases from the Great War. You are your parents' droppings, the remains of their discontent. Slothrop emerges from the wasteland without realizing the extent to which he is made of excrement. But Pynchon tells you how this culture turns life into plastic shit.

In 1925 Rilke wrote that "DUMMY LIFE" from America was replacing "the cared for animate houses, wines, apples" of Europe. Pynchon realized that what America was manufacturing best were plastic *people.* Imipolex-G, the polymer whose every fiber is capable of erection, is the sexiest cloth there is, the mystery stimulus that conditioned Infant Tyrone's erections. Were they measured against those of the foolproof, factory-tested polymer? Slothrop's father sold out his baby boy to a stimulus-response experiment in return for the money to send him to college. Pynchon is telling you that you are geared to excitement by synthetics, cast into your programming too soon to know what is happening, and too ignorant to realize that your father's love for your human possibilities was so meager that he was willing to plasticize you so that, alive or dead, you would get through Harvard.

Slothrop's love-hate affair with the V2 rocket is the paradigm of his conditioning, of your conditioning. The rocket outstrips sound; the noise of its coming arrives only after it has already exploded. Before you know what hit you, you are dead. This is Pynchon's most powerful symbol for the subliminal takeover of your mind. Every Infant Tyrone gets blasted by the violence of his parents' war with each other, by their rage toward him, by the anger of the Greta-Mother and the Blicero-Father who divide a child's physical and mental pain between them.

The American Oedipal situation is where you lose your valency, your attraction to everything, your enthusiasm for life. This is the chess game where the mother who would like to kill you and the father who controls you team up against the son who has to outwit them both. "Perilous Pop," Pynchon says, is the antagonist of every western, every comic strip. He is "every typical American teenager's own father, trying episode after episode to kill his son. And the kid knows it. Imagine that. So far he's managed to escape his father's daily little death plots—but nobody has said he has to keep escaping." Pop and his gang may not kill you, but they kill everything that makes life worth living. They steal the "Radiant Hour" from the day, steal life itself. Can anyone get it back?

Pynchon's rescue team is a catalogue of the kinds of people he feels this culture is producing, people bent on contemporary bliss: Myrtle the Miraculous is a wonder woman who hates people but adores the perfection of efficient machines. Marcel, the mechanical chess-player, is the ideal male, a robot tactician. Maximilian forsakes these fake humans, gets beyond male and female by allying himself with rhythms, all rhythms up to and including the cosmic. He's a fragmentation freak. Slothrop is the "glozing neuter" who cannot recognize himself as a man or a machine and whose fate is simply to run down ignorantly in the dimness of his vision. None of them finds the Radiant Hour.

Some of Pynchon's minor characters see the low-intensity light. Tchitcherine, a Russian drug dealer and spy, knows LSD is "The Lightning Latch, The Door That Opens You." It brings receptivity to the fact that only Death won the war. Tchitcherine sees a large white Finger addressing him, pointing out the rocket as the human soul, but Tchitcherine knows he will never see more than this twilight revelation. "He will miss the Light, but not the Finger."

Enzian, Tchitcherine's half-brother, is a half-black Herero who survived the great massacre of his tribe in South Africa with the serenity of a man who has passed through the worst. He suspects the rocket has a higher meaning. On an amphetamine high he sees it is connected with the purpose of human life. But speed does not bring the light.

The American and German streets are full of people looking for the Great Glow in the gold-star night with a pickup, the sensory flash. Franz Pokler, a German plastic man and engineer, gets high on incest fantasies that the endless supply of beautiful refugee children permits him to fulfill. Pynchon's Platz is full of anti-gravity forces—people popping pills, morning-glory seeds, or the "winerush" that rockets upward making "the woman screaming, the knife in your hand, your head down a toilet all unreal." The sensory trip is the New Dope. If you take it you see the profane light.

True Radiance begins with Byron the Bulb, the bright boy light bulb whose immortal beam screams "YOU'RE DEAD" in neon light. His real name is Thomas Pynchon, the writer who staked his immortality on being the man who illuminated the death at the heart of all experience. What happened to Pynchon between *V.*, the wildly sophisticated survival manual, and *Gravity's Rainbow*, the brilliant analysis of how you died? What happened to a writer who was not profane enough to take his own advice: "Keep cool, but care?" What challenged Pynchon's balanced gravity? Pynchon does not say. As his publisher puts it with terrific rightness, he keeps a "low profile."

But Pynchon offered a cautionary tale in *V.* in the saga of Fausto Majistral, who starts out to be a priest and poet. He gives up his detachment for love. He marries a woman he loves and has a daughter. He loses his faith, his work, his mind in an intimacy so disastrous it can only be described as world war. He writes to his daughter—born, like Pynchon, in 1938—"The bombs arrived with you, child." The birth of a child, like the profane nativity in *V.*, is death, the baby twisting like Godolphin's frozen spider monkey out of the Antarctic birth canal like a devil of need who shows up your love as a sham, your limitations as awesome. At the bottom of Fausto's mind is the memory of his father who was wrecked by war, of his mother who wanted to jump with her baby Fausto into the sea. Can death come from the willingness to breed life?

Pynchon has the clarity, the guts to see that what makes people kill and hate is not always a lover's rejection but a beloved's *responsiveness.* Given a choice between exaltation, or sensory amusement, too many people prefer the limited kick. What they cannot transcend is their gravity, the depression whose umbilical force binds them back to the stern down of father Death and pained mother Greta. Pynchon goes still further. He makes the most radical possible statement of the refusal to give up depression in Blicero's hatred of the lover whose youth and devotion are a challenge to his down. "O Gottfried, of course . . . you are beautiful to me but I'm dying . . . I want to get through it as honestly as I can, and your immortality rips at my heart—can't you see why I might want to destroy that, oh that stupid clarity in your eyes . . . when I see you so open, so ready to take my sickness in and shelter it, shelter it inside your own little ignorant love."

Love is a great reminder of limitations, of what you are not. The man who needs to dominate, control and crush spends his life hating his finite powers and trying to limit love. Blicero ties Gottfried up, puts him in a dog collar, forces him into a cycle of contempt and humiliation to bind up his feeling. Pynchon limits love to the S-M connection, where Blicero is the will to power and Gottfried the ability to love, each tying the other in knots. Through the novel Blicero is constructing the ultimate death-box, the V2 rocket fitted out for Gottfried, who will enter its nose cone wrapped in an Imipolex-G shroud. Gottfried, soaring on his love for Blicero, goes arching toward his death, while Blicero, Pynchon implies, dives straight into the flames of the rocket launch. Blicero is faithful to his gravity.

Pynchon affirms the loneliness unto death, makes this *Liebestod* a statement that there is *no* union, even in death. Blicero does not die with Gottfried, but rather makes sure each dies alone. The Radiant Hour for Pynchon is the hour of death, the fires of the V2 that liberate Gottfried and Blicero from the box of their own personalities, the shut trap of dominance and submission, into the molecular flow. What radiance Blicero and Gottfried achieve in their flaming deaths is the sparkle of illuminated filth, dirt purified by heat into the streaked glow of the rainbow that is not the sign of God's covenant with Noah, but the mark of Pynchon's imaginative connection to death.

Pynchon metaphorically traded his soul to the Devil for his own inviolability, his irrefutable alignment of all human endeavor on the axis of death. Pynchon embeds his own aggression in the patterning of a novel whose structure is a double cross. If the novel's theme is the rush of death through life symbolized by the trajectory of the V2 rocket, the novel's structure is a coordinate system in which the ordinate and abscissa coincide with Pynchon's *croix mystique,* the plus and minus of the soul. Christ and Antichrist, Good and Evil, Spirit and Flesh, Death and Life are the end points of the graph. But instead of being polar or opposite, they are all collaborators.

The spleen of the graph-cross is reflected in the novel's deceptive four parts. "Beyond the Zero" (I) states the novel's themes in terms of man's invasion of the highs and lows, the rational and irrational orders of science and mysticism for explanations of death. "Un Perm au Casino Hermann Goering" (II) stresses the constant, aggression, in every game of chance. "In the Zone" (III) moves deeper into the tunnels of the mind and of sex as Pynchon equates the SS insignia with Leibniz's double integral sign, the formula for calculating the densities of surfaces already known. "The Counterforce" (IV) is Pynchon's descent into the underground, the minus populations of dropouts, rebels and nuts who spend their lives in useless resistance against Death's Establishment. But it does not matter where you are in the novel. On Pynchon's coordinate system, the only thing that can be calculated is how close you are to death.

Pynchon always brings you to zero. His seduction into order, patterns, dualities is meant to increase frustration. Those who are aroused by order and compassion are the first to go. Pirate Prentice is an English intelligence officer who makes sweet banana breakfasts for his friends, whose gift is for assuming the burden of other people's anxieties. He does his job after World War I by allaying the fear of a Balkan Armageddon. He has no idea that his gifts help Death by keeping England diverted while the Devil does his work in Germany. Leibniz said, "Le présent est chargé du passé et gros de l'avenir." Pynchon's karma has it that kindness, sweet dreams and banana breakfasts turn to excrement and ashes. On page one of *Gravity's Rainbow* it may be grateful Death who is dreaming of welcoming Prentice to hell in velvet luxury. Prentice does not know he's damned until page 537, when he gets a tour of hell. Pynchon lets you, the reader, discover only on the novel's last page that his rocket is aimed at you.

Pynchon wrote of the dead hero of *The Crying of Lot 49:* "He might have tried to survive death . . . as a pure conspiracy against someone he loved. Would that breed of perversity prove at last too keen to be stunned even by death, had a plot finally been devised too elaborate for the dark Angel . . . ?" Pynchon survives his destructiveness by turning it into a novel too complex to escape. He is the artist of tortured entrapment and limitation. He did his bit to limit life by boxing all experience into one either/or: the mechanical symbiosis of *V.* or no life at all. But Pynchon went still further in ironically affirming limitation as the sole purpose of existence. Given our destructiveness, our need to kill, to sully life, our mission on earth, Pynchon concludes in *Gravity's Rainbow,* must be to celebrate the Devil. "Our mission is to promote death."

Kepler conceptualized gravity as the Holy Ghost for "physical and metaphysical reasons." It was God's love, he thought, that swept the planets around the sun, kept them in place, favoring earth with its green corona. Pynchon conceptualizes gravity as a parabolic rainbow, also for physical and metaphysical reasons. The rainbow is Death's hate, Death's grimace, the tragic mask of the heavens pulled down forever in one inviolable affirmation of depression. And in his myth of himself as Death Incarnate, Pynchon transcends his limitations, puts himself beyond the pale of human pain and cruelty. He allies himself with the ultimate aggressor, the impersonal force of the Entropy God. In the throes of his pessimism, by *force* of his pessimism, Pynchon is still pursuing his own invulnerability.

The dream of vulnerability conquered is the dream of the age. Pynchon has an intense sensitivity to the evil the dream contains, an analytic brilliance at extracting the villainy behind every smile, a stunning accuracy about what is wrong with emotional life in this culture now. He seems pressed between the dream and his contempt for it. His pain, his vulnerability, his great and ruined expectations keep breaking through his intellectual hardness. Pynchon's own refusal to stop demanding that life be perfect caring or perfect emotionlessness, his inability to stop making conditions on life that life cannot fulfill, and his own pain weld into a pessimism so unassailable it becomes an argument against pessimism. Pynchon's indictment of every human impulse is his crucifixion on the modern dream. It is so intense that it has a cautionary force against gravity. Pynchon's most eloquent moral is himself.

Pynchon is a genius as the poor devil who went beyond the grave to anatomize the remains of the modern soul. Is it courage or sheer perversity that makes Pynchon take to the very edge the tendencies of a period marked by war, by megacartels, by sexual anger? Pynchon is an illuminator of faults. He shows in the once-graceful male and female principles, Reason and Romance, the latent evils that surface as the historical forces that shape our physical environment and provide us with dreams, those images of ourselves and each other that produce the aggressions of technology and of sex. Industry and expressionist art, chemistry and comic books, the multinational corporation and the lonely control addict are where history and the individual intersect.

Pynchon mines the dark side of our dreams of order, the forbidden land where wholeness and control really mean exploiting others. He mistrusts answer-men; he mistrusts reason; he mistrusts himself. He is brave enough to refuse to give things false names. When his Mister Information tries to call Pain City (a city more like London or New York than Rilke's *Leidstadt*) Happyville, Pynchon snickers. He mocks even the sentimentality Rilke lavished on his innocent boy who, in the tenth Duino Elegy, climbs the mountain of primal pain as Rilke affirms the river of joy at its base and the happiness of the boy's eventual fall toward infinite being. Pynchon tickles Gottfried toward his macabre death by wrapping him in an erectile Imipolex-G shroud. The best Gottfried will get is a plastic caress and then cessation. Pynchon may be in love with cessation. But he shows that in our love of falling, of letting go the burden of ourselves, we ape dead matter. By his insistence on the harshest possible vision of both our self-protectiveness and our self-erasure he blocks the enthrallment to death.

Pynchon has put together the emotional, cultural and historical preoccupations of his generation with a brilliance and depth that match what Thomas Mann did for the prewar world in *The Magic Mountain*, that equal James Joyce's compendium of his time in *Ulysses*. He plays Beethoven to Rilke's Schubert, developing from Rilke's encapsulated emotional statements operative definitions of the nature of science, thought and civilization. Pynchon may not believe in salvation, but in the imaginative power of *Gravity's Rainbow*, this mournful genius showed that the Devil could be a Messiah.

X

THE PURPOSIVE IMAGINATION

For I kept my heart from assenting to anything, fearing to fall headlong; but by hanging in suspense I was the worse killed.

St. Augustine, *The Confessions*

Holistic and anarchic fiction may be seen as a protest against the human condition both for its finality, thanks to death, and its rigidity, thanks to convention. Although our novels are not, in the main, novels of ideas, they are concerned in simple, concrete ways with contesting the origins and ends of human activity. Questions of morals and metaphysics could be pressed by novelists of the nineteenth century into the prevailing dramas of American puritanism and of the disappearing wilderness. Writers of the 1930s could rely on the evils produced by the industrial revolution to widen a sense of spiritual depression. However, the social facts which once enlarged and reinforced the sense of individual injustice seemed, for postwar novelists, to have been made obsolete. Ironically the relative success of both American capitalism and the labor movement seem to have virtually impelled the novelist to look inward to continue the inquiry into life. In our middle-class, mainstream fiction, our spiritual condition is imprinted in a revolution of sensibility, boring into questions of character, meaning, and significance.

The revolution in our literature is largely not an argument with the real world. It proceeds by conciliation or appropriation rather than by insurrection. Postwar novels reflect a vision of incorporation and accommodation as revolutionary acts. The widening wish to include all possibilities, to reconcile all opposites, is a weapon against the power of any one idea or value. In the interest of devaluing the sense of life as a struggle between right and wrong, good and evil, conquerors and victims, all distinctions between truth and falsehood, fiction and fact, inner and outer reality are destroyed.

To see reality as only a point of view is to apply to life the values of art, of artfulness. This conversion of the world into a subjective circus may be one outcome of our intellectual heritage. In the early revolt against religious rationalism. Nietzsche argued that distinctions between good and evil were only forms of psychological accommodation, ways of rationalizing self-hatred and weakness, and Freud showed the infantile, subjective origins of religious impulses. For neither did the pliability of reason and language lead to a judgment against the value of making judgments of truth, value or meaning or against the value of reason.

But in many postwar novels intelligence is judged either as a diversion from the real world or an instrument for gaining personal or political power. Thought is a funhouse full of dead ends for Saul Bellow's intellectuals, whose ideas provide relief from personal turmoil but no solutions. Pynchon's Herbert Stencil is meant to be prototypical of a man looking for an answer he does not wish to find. The intellectual in the novel has frequently learned to use ideas as strategies against recognition. Because reason is pliable it is also judged to be a manipulative tool. All ideas seem merely propaganda for the production of myths about the individual (Barth), for the exploitation of others (Pynchon) or for consoling deceptions (Vonnegut). In a spirit of intellectual mistrust, Thomas Pynchon, who may have the deepest analytic intelligence of any novelist of his generation, attributes to scientific curiosity and the spirit of inquiry most of human destructiveness. In this contemporary mind, reason is being returned to the realm of theology and judged as evil.

Anarchic and holistic fiction carry out a relentless torture of reason, as well as a reexamination of values. In postwar novels, rationalism, self-examination, self-knowledge and mystical insight may all be devalued. The refusal of a fictional character to make distinctions, to believe the world may be other than what he says it is, is a tool for a kind of self-liberation. It brings freedom from the idea that human nature is fixed by death, by the determinisms of psychology, heredity or environment. Envisioned as a series of experiences, a scenario where anything can happen, life and death may be seen as reversible encounters with no objective order or rules. The seductive illusions and deceptions of art thus form a core of revolutionary values.

Anarchic fiction advances a kind of mindlessness as a virtue, finding in meaninglessness, fragmentation, memorylessness, and release from personality ways out of destructiveness and self-destructiveness. Holistic fiction takes an opposite course, increasing the pliability of reason to the will. It uses intelligence as a form of propaganda, a hortatory device that calls for control, order, and progress not through insight but through expertise. It speaks through technological images of robots and computers, through the dream of other worlds where destructiveness and self-destructiveness can be switched off. It seems to caricature William James's definition of ideas as solutions to experiential problems. "Thought," James said, "is what pays."

Both anarchic and holistic fiction deal with the ancient subject of art, the conflict between good and evil, by inverting the meaning of the words good and evil. What is good in anarchic fiction is to disappear, to regress, to revert to childhood, to find a door into a sense of life as blessedly pointless, without guilt or responsibility, and into a sense of death as an endlessly proceeding, reversible molecular flow. Holistic novels find good in the opposite task of developing techniques for pulling life together on a mechanical model. The holistic spirit performs an act of retrieval, salvaging fragmentary lives for participation in some operational scheme that is often the closest it can come to life.

Holistic and anarchic fiction are consequences of the subjective world view. Because all is illusion in the novel, it is an excellent vehicle for the image of life as a fantasy. Descartes' assetion "I think, therefore I am," is distorted into the faith that character, ethical standards and values are subjective, questionable, tentative ideas in a single mind. Where the intellectual effort of other times has been to try to salvage some hard and permanent truths from the world of confusion, the literature of our time reflects a flirtation with the state of being divided, uncertain, in revolt, in retreat. Far from being simply problematic literature, it reflects the extent to which the problematic has been eroticized, lightened, made a pursuit capable of giving pleasure. In fiction, the dominant intellectual and perhaps moral style is paradox, the sharp faith that both a statement and its opposite may be true. Although our novels may not tell the *whole* truth about us—we are more alive, more loving, more politically sane than our novels suggest—they tell the truth about an American revolt that operates by conciliation.

What Christian heritage exists in mainstream postwar literature may be evident in its tendency to encompass all distinctions, its effort to bless all human striving and to consider every character as salvageable. Many postwar novelists are moralists who use the absence of precise values to make a point about the impossibility of separating good and evil in a single human heart. The moralism to be found in Barth, in Pynchon, in O'Connor, in Oates and in others shows a sense of the attrition of the distinction between god and devil, good and evil, exaltation and misery. It forces opposite values into a paradoxical relation. Who can tell what is good or right in O'Connor's stories of sacred Misfits, of the sinful and the righteous burning together in heaven, or in Oates's tales of deprivation and sacrificial love, or in the losing battle for life that, in *Gravity's Rainbow,* leaves the laws of entropy in control of human activity? In the consuming moral confusion, good and evil, life and death, penetrate each other. The devil is given his due as a charmer, evil is credited with power to enchant. It may be that from all this confusion, only confusion is affirmed. Yet to me the sophisticated moral ironies in our novels point toward the necessity to recognize that even such "revolutionary" subjective, paradoxical values are manipulations of the real world. Camus defined nihilism as partly "the inability to believe in what is, to see what is happening, and to live life as it is offered." The moral paradoxes in fiction seem to point out not only the need for acceptance of the universe of emotion and passion, but also for faith in and accuracy about the real world and for the emergence of a more precise morality inseparable from lucidity.

The novel mirrors a revolution that is not material but spiritual. Its goal is consolation. Based on a sense of the lasting imperfection of character, it remains perpetually open to the dream of change for the better. It reflects a time too acutely aware of power and submission and too addicted to the image of all reality as subjective. The novel entertains by being entirely perspective and point of view, by offering an array of possibilities which have equal aesthetic weight. In the novel there is no difference between the world and a character's vision of the world, no necessary difference between the merits of any particular point of view. Consider the story of Li Po. "Li Po sleeps. An assassin creeps in and raises a knife above his throat. Li Po wakes, sees the knife, and smiles. Who has the knife?"

Anarchic fiction uses comparable passivity, or comparable instances in which a character is overwhelmed by life, as an entertainment. It reassures not by challenging the situation, but by contesting how one deals with it. Anarchic fiction accepts the smile as a solution, accepts the existence of the adversary and deals with him by manipulating how he is seen, how one feels about him. It inverts standard values to justify a sense of submission. Disintegrative themes express a sense of universal impotence. They reject the holistic view of life that stresses performance and aggressive confrontation. Anarchic fiction often magnifies the odds against the hero by making his adversary too large for anyone to slay.

Who could kill the protean dragon of modern times? In Vonnegut, the enemy is fate, the universal programming for suffering, or World War II, or a magic formula that ends the world. It spurs the hero to decide that the best he can do is look the other way, admit his helplessness, and make a religion out of his inability to change his life. In Heinlein's pop science fiction, fear and sexual repression make the crowd kill the Martian Messiah who preaches nonaggression. The toughness of the adult world of tigers does in Brautigan's dreamy heroes. The rages of the world are too great for these characters to withstand, so they simply smile, in the hope of enduring. Denying the "I," seeing it as a fragment, provides protection against destructiveness and self-destructiveness. As Odysseus outwitted the Cyclops by claiming he was Noman, so do many modern characters avoid painful confrontations by claiming, in effect, that they are not there. In breaking the self apart, this fiction reflects a revolt against the selves the culture demands.

Anarchic fiction reflects a world perceived in terms of sufferers, victims whose victimization and oppression are solved either by being placed beyond their control or by being elevated to a form of fulfillment. Bellow's Herzog is a victim but his victimization is inevitable and purposive; it exists to perpetuate his youth and to provide a peculiar sense of security. Vonnegut's Billy Pilgrim is a sufferer who finds on the planet Tralfamadore a vantage point from which to perceive his unhappiness as irrelevant. Vonnegut has a comic vision, Bellow offers a form of tragedy in the felt misery of Wilky in *Seize the Day* or the confusion of Leventhal in *The Victim*, in all the characters who curiously find release in awareness of helpless inadequacy.

Most postwar novels search for a vantage point from which life will not appear so irremediably painful. Anarchic values (the emphasis on dismantling character, through time-space travel, through regression, through a sense of the reversibility of time) function in holistic novels as temporary escapes from the combativeness of life. In holistic books where life is put together in terms of war, disintegrative techniques are shock absorbers. In William Burroughs' and Hubert Selby's fiction, the body is the arena for social and class warfare; the rape, torture, and sexual degradation of men are made to appear both universal and incurable. In novels in which people get themselves together only to explode, collapsing personalities, people who are anonymous body parts, or the use of interchangeable characters provide relief. Such dissolution, like addiction to the pain-killer heroin, permits the hero to avoid personal recognition of anger and pain while still seeing life in terms of power relations.

Pynchon most brilliantly handles the alternation of disintegrative impulses with forces for order. In *V.* the emotional core of people is contained, but never fully controlled, by the desire for order and peace. The movement of history operates as an alternating play of aggression and containment, of id and ego, war and diplomacy, anarchic and holistic impulses. In *Gravity's Rainbow* the forces for order eventually lose out to the forces for dispersion. The anarchic solution advances by retrogression and primitivism, by return to anonymous impulses, always bucking the demands of cultural facts and institutions. Anarchic fiction embraces impersonality as the answer to being treated impersonally by a universally malevolent fate. The molecular flow of life dissolves all, but includes all. It releases the hero from memory, self-examination, all distinction and all comparisons.

Holistic fiction virtually welds the biologic imperative to live to a behavioral conception of personality. Its techniques are parody, pastiche, plot; its concerns are with mechanisms for endurance in a time of the abrasion and dispersal of the self. To be or not to be is a frequent explicit subject in holistic fiction.

John Barth, John Gardner and Robert Coover are fascinated by the line between what keeps people going and what makes them fall apart or become outcasts. In Barth's novels, the hero resorts to some external force—the computer, the logical construct, or role-playing—to keep himself going. In Coover's fine novel, *The Universal Baseball Association, J. Henry Waugh, Prop.*, a lonely accountant draws companionship and sexual power from his fantasy life as creator of the Universal Baseball Association and his identification with its star pitcher. In Barth's and Gardner's novels, the hero survives through becoming a replica of a mythic figure. The pastiche character whose being and acts are paste-ups from other stories achieves not only detachment from his inner confusion but a large certainty and authority through his analogous existence. By playing roles he draws from the pipeline of old narratives purposiveness and strength; he incorporates a power and wholeness he does not himself innately possess. The act of creation is virtually synonymous, for the character, with the act of appropriation.

In a culture in which work is subordinated to production, the creative imagination practices a utilitarianism of its own. Nihilism in holistic novels can be called, as it is by Barth, cheerful because it is also useful. Holistic novels are filled with controls on aggression and destructiveness. Pynchon, in *V.*, develops the inanimate as both a nightmare and the solution to human vulnerability. From Eigenvalue the dentist who prefers metal teeth (they have no pulpy soft center), to Benny Profane's fantasy of a mechanical woman (you could call her Violet) who, if she gave you trouble, could be cured by replacing her parts, to Rachel whose first lover appears to have been her MG, the quest for an inhuman other world, unperishable and painless, is ironic and ever present.

John Barth's unsentimental awareness of the problems of intimacy offers comic visions of how to avoid it. In *Giles Goat-Boy,* for example, a masochistic girl and a boy who has been raised among goats enter the belly of a computer to discover the mystery of everything. They make love in the machine. The girl finds release from her masochism, the boy from his animal insensitivity. The control on her self-destructiveness and his destructiveness is the monitor robot. Outside the computer they never achieve that happiness again.

Holistic fiction tries to structure sexual relations, looking for the computerized control on aggression. As absolutely as disintegrative novels advocate blurring external experience, holistic fiction tries to sharpen the human situation toward simple, unmistakable behavior or roles. Holistic fiction places an extraordinary value on clear perception. It stresses a kind of realism that is also a means of seizing control. Barth's Todd Andrews chronicles all the events that led up to his decision not to kill himself. To list, to count is made to seem the same as knowing and solving one's world. *Lookout Cartridge,* a complex, cerebral novel by Joseph McElroy, uses the movie camera as an image of consciousness conceived of not as an analyzer of the meaning of experience, but as a recorder. The novel not only offers a point of view on reality, but manipulates the mind by shifting perspectives of the real world.

Novels of aggression set in the postwar period magnify the tendency to perceive life in terms of external manipulatable situations. World War II is continuously an image for the uncontrollable. Holistic, political novels set in the postwar period avoid it and deal with smaller adversaries. Norman Mailer's *The Deer Park,* Robert Stone's *Hall of Mirrors* and Ken Kesey's *One Flew Over the Cuckoo's Nest* perceive the world in terms of a cold war standoff, a pause which gives the opportunity to see the rebel and the tyrant as alter egos. The debunking, revisionist political eye playing over America in these novels is largely cynical about American institutions and character. The rebels in these novels are so like the system they hate that there is little value or meaning to their political revolt. The subject of the novel turns from the world, or the nation, to the situation of man in revolt, the feelings of the man who is divided, yearning for some unattainable freedom but caught by a thirst for the system's blessings. His own conflicts and appetites become more the issue than the political climate. Inevitably sex is the index of his revolt.

The primacy of relations between the sexes as a subject in postwar fiction reshaped the literary sense of maleness, of revolution, and of heroism. Freud observed that much behavior that was not sexual was sexually motivated. In the postwar novel the opposite is true. A great deal of sexual behavior is not sexually motivated. The drive of the male character is frequently power or acquisition, not even lust. The oversimplification of sex into a destructive or self-destructive addiction in Burroughs and Selby reflects an unrelieved kind of male combativeness. The much-maligned male chauvinist characters of Mailer use women as the territory to be conquered in their quest for a vision of themselves as heroes. Heller's Bill Slocum and Lois Gould's transsexual B.G. use sex to reassure themselves that they are still alive. The drive is not lust but anger and power, a fact which may explain why so many sexual heroes remain so persistently unsatisfied.

Gardner, Bellow, Updike and sometimes Mailer move toward a more complex vision of sex as the organizing principle of a man's life. In Gardner's romantic, pastoral novels, love for a wild but interesting woman is a disruptive, force that can destroy a man. A woman can even be a worthy adversary, a witch like Medeia who can foil a rationalist Jason, or a flinty, opinionated old lady as fierce as her flinty brother. Mailer's men are torn between submissive women they hope will arouse their tenderness and aggressive women who may keep the man's aggression in check. Bellow and Updike show men sentenced to play the Oedipal drama for life. Bellow's men have no conscious awareness of or desire to articulate the reason for their persistent unhappiness with women. Updike's men cannot change their ambivalence over being both fathers and sons, but Updike's unsentimental intelligence achieves a clear statement of their problems. It is Updike who provides the fullest revelation of a man's hesitancies in an ongoing marriage, of the subtleties of sexual need and anger between a husband and wife, and of the opposite pulls of freedom and responsibility.

It is not misogyny that figures most prominently in such novels of sexual relations but frustrated dreams of achievement and self-love, and of release from perpetual doubt. Updike's men find momentary rebirth in sex, but the emotion which frees them from sex as family entanglement ironically releases them into the additional entanglements their affairs create. As Mailer's Eitel puts it, "The purpose of freedom was to find love but when one had found love, all one could want was freedom again." Bellow's Herzog sentimentalizes his mother and does not look deeply into his life with women. Woman becomes the symbol for the hero's larger connection to life: for the adventurer, she is another territory; for the religious man, the means of exaltation; for the nostalgic man, the catalyst for looking back.

Woman is the symbol of *disconnection* in novels where sex is persistently another dead end. Tyrone Slothrop in *Gravity's Rainbow* cannot help the fact that his love may be the death of his pickups. He is romantically looking for a woman who will unlock the mystery of his anger and release him from it, but finds only women who are exactly like himself. Robert Stone's Rheinhardt is initially fascinated by a woman who appears as hardened and scarred as he is. But the woman cannot duplicate his aloofness, and he feels little protectiveness toward her. In *V.* Benny Profane wishes for a mechanical woman whose parts can be replaced when they give him trouble. In *Dog Soldiers,* Converse hooks his mistress to him on heroin. The sexual bond is the connection used as an image for emotional disconnection. Sexually saturated postwar fiction seems to be pointing toward the decline of love as a subject. The prototypical hero of holistic fiction does not see being a husband or father as the mark of his wholeness as a man. He pulls himself together by sexual acts which are ultimately aimed at separating him from the world around him. The amount of energy directed against women is often only an indication of his pervasive need for discontinuity and the value he places on alienation.

Mark Twain charged that the passionless, super-civilized virgins in James Fenimore Cooper's novels had never existed anywhere and were making it impossible to describe real women. But now the situation of women in the novel is not so different than that of men. A feminization of the male hero has occurred. The narrow sphere of activity, the family, encircles many male characters, men who can confront disruptions in society only as represented by their irate sisters and wives. The sense of vulnerability distinctive to women in nineteenth-century novels is almost universally shared by current male heroes. Women have been taught vulnerability as a means of gaining male protectiveness and other benefits of civilization. Now the male hero also accepts and becomes resigned to his situation in order to resolve tension. In feminist novels the heroine is struggling toward a dominant position. Yet in novels by both men and women the heroine's desire to change is reactive, caused by disillusionment rather than conviction or ideology. The sad sack heroines who have been bred for intimacy and have found it wanting have presumably become disillusioned by marrying men like the heroes of Heller, Mailer, Updike, Bellow, Stone and Pynchon, who are bent on avoiding intimacy.

There is much greater variety in the way male characters deal with the degree of anger between the sexes. The man who finds peace in fragmentation resigns himself and numbs himself; the one who sees life in terms of power can usually find a woman to subdue. More recent heroes like Rheinhardt or Profane keep aloof from involvements. The heroines of Oates who submit ecstatically or masochistically, or of O'Connor who strike back, or of Didion who go through terrible experiences narcotized, represent exceptions. In other novels, however much the heroine admires the ability to be accepting, exploitative or aloof, she has not succeeded in learning how to be any of these effectively. She remains curiously undefended against the onslaught of experience. Such heroines remain apart from the mechanistic drift of recent fiction, although the holistic world view stressing control and coolness remains an ideal.

Few postwar male novelists have achieved portraits of women comparable in depth to those of Henry James or even of Theodore Dreiser in *Sister Carrie.* There has been no postwar novel by a woman as perceptive about male strengths and vulnerabilities and as revelatory of male destructiveness as those by Barth, Pynchon, Updike or Stone. It is men who have, intentionally or not, revealed most about the rich perversity of the male mind. I know of no male character created by a woman who is as unlovable as Heller's Bill Slocum or as involuted as Bellow's Herzog. The cutting, cruel betrayer in Lois Gould's *Such Good Friends,* the macho drifter in *Looking For Mr. Goodbar,* or the long-suffering suicidal observer of his wife's infidelities in *Run River* pale beside the male novelist's cast of male characters who are sexual angries or withholders, who may be narcissistically committed to themselves alone, unaware they have any desire to wound women, who consciously enjoy ripping up a woman's mind, or who expend most of their energy in a gentlemanly effort to control that impulse. Men apparently have fewer illusions about themselves than women have about them.

Idealism about men and love, and even faith in romantic love, survive in feminist fiction as comedy, as irony, as the stuff of cautionary tales, but always as an unavoidable emotion. In the novel disillusionment with men is a fact difficult to face and once faced, apocalyptic in its consequences. The debunking of women's traditional lives, and the refusal to separate the problems of being alive from the problem of being a woman, often lead the feminist writer to extreme statements about the impossibility of happiness as a woman with men. Extremism is always characteristic of revolutionary art, but in a revolution of sex it has a particularly painful force, underlining the extent to which each sex is so engrossed in the drama of its own vulnerabilities that it loses sight of the other.

Fiction exploits the breakdown of social and sexual roles and expectations as a solution to tensions. In the novel the peculiar ethics of meaninglessness fill the vacuum once occupied by family values and old sexual stereotypes. The sense of the pointlessness of human connections in the novel is meant to have the effect of armor. Images of life as a floating opera whose acts one can never wholly observe, of history as lessons no one learns offer shelter from the perception of civilization as a tragic love story. If the pieces of experience are connected they may unfold, as Pynchon describes in *The Crying of Lot 49*, as a vision of life in which men and women are pawns in a revenge play written by fate.

The meaning of work is inevitably under fire in a culture where value is measured by utility and status. From Brautigan's little corporal who works hard only to discover that no matter what he does he cannot rise as high as the rich kids in his class, to Coover's accountant who drowses over his books by day but comes alive keeping records of the imaginary Universal Baseball Association by night, from Stone's Rheinhardt who gives up the clarinet he loves for the small pleasure of exploiting the system, to Slocum who performs a well-paying but useless job, there is little sense of the excitement or pleasure of doing something well.

The attempt to defeat the problematic nature of love and work, to put the world at bay, takes paradoxical turns. In anarchic fiction the problems are magnified, the value of human effort is minimized, and the attempt to shape life ceases. Conversely, holistic fiction minimizes the problems blocking the individual and offers operational cures. Both reflect the attempt to shift the terms of human discourse onto different levels. Anarchic fiction and holistic fiction represent distortions of the human situation which lead to different ends, the one toward the impersonality of mysticism, the other toward the impersonality of technology. Postwar fiction reflects a spirit of divergence in our time. It shows the deification of the uncontrollable, the random, the fragmentary, and the deification of the performer, of man as a success machine. Both bear witness to our fascination with vulnerability and power.

In the real world the values reflected in anarchic fiction have produced terrible results. Charles Manson perceived that the person who cannot tell the difference between one feeling or idea or Messiah and another could be a murder weapon. When Susan Atkins, who believed Manson was Christ, said about killing the pregnant Sharon Tate, "You have to have a real love in your heart to do this for people," she revealed not only the extent of her own pathology but the most horrible result of the ethics of meaninglessness. Nothingness inevitably comes from a belief that everything can be true, or right.

In fiction the anarchic impulse is largely benign and defensive, concerned with dissolving painful power relations and political, intellectual or sexual differences. But even its most brilliant occasional practitioner, Thomas Pynchon, is a moralist who sees it as having won out over any human attempt to integrate, preserve or organize experience. He portrays those of us born in World War II as a generation stamped by self-destruction and destruction and only deluding itself in thinking it can tell one from the other. Pynchon sees both our attempt to organize reality and our commitment to letting go as equally damning, equally signs of human abdication to the formal laws of destruction, as kinds of political, emotional and sexual entropy.

Holistic fiction performs an act of retrieval, of creation, through affirming the value of life. The world of dispersion of fragments, of process without a point is pulled together in this modern Apollonian art in the technological images that fill our lives. It makes out of the impermanence and flux of modern life, by means of will and intelligence alone, an affirmation of remaining human unity.

Postwar fiction has achieved an extraordinary revolution against the literary constraints imposed by the dictates of romanticism and naturalism. It has redirected the inflated heroic dimension of the former and the rigid determinism of the latter to achieve characters who more closely approximate the stature of real people and who seek for the greatest possible flexibility in shaping their lives. Perhaps it should correct its correctives, should show more of the heroic love or political optimism that are present in people. Yet it has been brave enough to tell unpopular truths and dared to show that even insoluble problems, even journeys down dead-end roads, can be sources of entertainment and enlightenment.

Postwar fiction has put the problematic in perspective. It bears witness to the adaptability of people. In a time of emotional trauma and reappraisal, it seems bent on showing that even a labyrinth can become a home. Now the unity of fiction lies in the disposition to discover what can happen next and to attempt to encompass all possibilities, all subjects, all emotions, all ideas unselectively. In this odd, heartbreaking democracy of experience, perhaps American innocence may breathe again, breathe with the compassion St. Augustine expressed when, faced with the need to explain malice and failure in the world, he wrote: "And it was manifest unto me, that those things are good, which yet are corrupted, which neither were they sovereignly good, nor unless they were good, could be corrupted."

The protest against pain, like the revolution in morals and values, will no doubt continue in the pages of novels. To deserve to succeed in the real world, such a revolution will have to supply ideals that arouse constructive political and social effort. By confronting both the massiveness of our problems and our inventive energy, postwar fiction shows the continuing purposiveness of the imagination. It performs a service by casting shadows as well as light, by making us see with a more powerful and compassionate lucidity.

INDEX